PRACTICAL HANDBOOK

OF

CURVE FITTING

EDITOR-IN-CHIEF:
SANDRA LACH ARLINGHAUS, Ph.D.
Institute of Mathematical Geography
and
The University of Michigan

ASSOCIATE EDITORS:
William C. Arlinghaus, Ph.D., Lawrence Technological University
William D. Drake, Ph.D., The University of Michigan
John D. Nystuen, Ph.D., The University of Michigan

CRC Press
Taylor & Francis Group
Boca Raton London New York

CRC Press is an imprint of the
Taylor & Francis Group, an **informa** business

COVER PLATE
The cover plate captures fitting data to curves and curves to maps. The data base used was from the World Resources Institute; a table from that data base was exported to Lotus 1-2-3, release 2.3 where the curves were fit to the data, and the graph made from the data. The 1-2-3 file was then imported into MapInfo for the PC and the data was further analyzed and graphed.

FIGURES 5.6 to 5.8d
These figures are reprinted with permission of the American Geographical Society; see figure captions and associated reference.

FIGURES 7.1 to 7.6
These figures are reprinted with permission of *Economic Geography;* see figure captions and associated reference.

MAPS, GRAPHS, DATABASE, AND TEXT
These maps were made by S. Arlinghaus using MapInfo (licensed to W. Drake/CSF) and Atlas GIS (licensed to W. Drake/CSF) on a 486 Personal Computer with 16 Megabytes of RAM and a 240 Megabyte hard drive. The graphs were drawn using Lotus 1-2-3 release 2.3 (licensed to W. Drake/CSF). The databases used were formulated from electronic files of public domain tables from the Stars database of The World Bank and the WRD database of the World Resources Institute. The text was prepared by her on the same computer using MicroSoft Word for Windows, version 1.1 (licensed to W. Drake/CSF).

CRC Press
Taylor & Francis Group
6000 Broken Sound Parkway NW, Suite 300
Boca Raton, FL 33487-2742

First issued in paperback 2020

© 1994 by Taylor & Francis Group, LLC
CRC Press is an imprint of Taylor & Francis Group, an Informa business

No claim to original U.S. Government works

ISBN 13: 978-0-367-57980-7 (pbk)
ISBN 13: 978-0-8493-0143-8 (hbk)

Visit the Taylor & Francis Web site at
http://www.taylorandfrancis.com

and the CRC Press Web site at
http://www.crcpress.com

Library of Congress Cataloging-in-Publication Data
Practical handbook of curve fitting / editor-in-chief, Sandra Lach
   Arlinghaus ; associate editors, William C. Arlinghaus, William D.
   Drake, John D. Nystuen.
       p.   cm.
   Includes bibliographical references and index.
   ISBN 0-8493-0143-4
   1. Curve fitting.   I. Arlinghaus, Sandra L. (Sandra Lach)
   QA297.6.P73   1994
   001.4'225--dc20   94-10118
                                                            CIP

Library of Congress Card Number 94-10118

PREFACE

This handbook is a reference work demonstrating how to analyze World data bases and graph and map the results. Default settings in software packages can produce attractive graphs of data imported into the software. Often, however, the default graph has no equation associated with it and cannot therefore be used as a tool for further analysis or projection of the data. The same software can often be used to generate curves from equations. The reader is shown directly, and in a series of steps, how to fit curves to data using Lotus 1-2-3. There are traditional unbounded curve fitting techniques--lines of least squares, exponentials, logistic curves, and Gompertz curves. There is the bounded curve fitting technique of cubic spline interpolation. Beyond these, there is a detailed application of Feigenbaum's graphical analysis from chaos theory, and there is a hint as to how fractal geometry might come into play.

Curve fitting algorithms take on new life when they are actually used on real-world data. We use them in numerous worked examples drawn from electronic data bases of public domain information from the Stars data base of The World Bank and from the WRD data base of the World Resources Institute. The applications are current and they reflect a state-of-the art interest in the human dimensions of global change.

Useful Handbook Features:

1. The comprehensive character of the book draws on a mix of the academic and the international development's community approaches to data involving global change.

2. Step-by-step instruction on curve fitting.

3. Illustrations of how to use maps to highlight the spatial dimension of global data.

4. Case studies in a global change context using global databases.

5. Indices to the figure captions and table titles. Often what readers remember, especially in a book with a great many maps, figures, and tables, is one particular visual display. These indices offer readers seeking one picture they remember an easy-to-use method to find it.

As is necessarily the case with this sort of endeavor, it needs immediate and continuous updating--as this first edition goes to press, a second is in the works. If you find something that you wish to share with us for the next edition--new material, corrections, or whatever-- please communicate with:

William C. Arlinghaus
ATTN: CRC Handbooks
Department of Mathematics and Computer Science
Lawrence Technological University
21000 West Ten Mile Road
Southfield, MI 48075
Arlinghaus@LTUVAX

Thank you in advance; we have tried to be careful and to contribute something that is useful and different; in the end, despite all the care of the many who have generously offered time, effort, and advice, the blame for errors, omissions, or poor judgment must rest with the Editor-In-Chief, alone.

Sandra Lach Arlinghaus, Ann Arbor, MI   November 23, 1993.

ABOUT THE EDITORS

This book was written primarily by the Editor-In-Chief, Sandra Lach Arlinghaus. The role of the board of Associate Editors was to review and critique the material and guide the direction of the document: in terms of general scope, databases used, themes to investigate, graphic presentation, and accuracy of detail.

The set of five editors has combined experience of well over 100 years in various disciplines and activities related to curve fitting, including an extensive list of publications and invited/contributed lectures, at the professional level. Areas in which they have professional expertise include: geography (mathematical, spatial analysis, cartography, GIS), regional science, urban planning, intelligent vehicle highway systems, transportation planning, natural resources, public health, population planning, operations research, engineering, mathematics (graph theory, combinatorics, modern algebra, non-Euclidean and Euclidean geometry), computer science (theoretical), computer science in the automotive industry, photography, and field study in the international arena.

Sandra Lach Arlinghaus, Ph.D. Geography, University of Michigan.
Founding Director, Institute of Mathematical Geography.
Adjunct Professor,
School of Natural Resources and the Environment
University of Michigan

William C. Arlinghaus, Ph.D. Mathematics, Wayne State University.
Professor and Chairman,
Department of Mathematics and Computer Science,
Lawrence Technological University

William D. Drake, Ph.D. Operations Research, University of Michigan.
Professor,
School of Natural Resources and the Environment, School of Public Health, and, College of Architecture and Urban Planning,
University of Michigan.

John D. Nystuen, Ph.D. Geography, University of Washington.
Professor,
Geography and Urban Planning,
College of Architecture and Urban Planning,
University of Michigan.

The research for this book was drawn from the field experience of the editors and from background reading listed in the references section. We owe a particular debt to numerous colleagues who have gone before us; we hope that this handbook will supplement their fine materials and be of use to students, researchers, business people, and others in a variety of practical contexts.

ACKNOWLEDGMENTS

Over the past two years Editor Drake invited S. Arlinghaus to co-teach his graduate seminar on transition theory at the University of Michigan. At first an experimental course, School of Natural Resources and the Environment, it became in the next year a regular offering in both the School of Natural Resources cross-listed in the School of Public Health. In these courses, students interested in global change worked with each other and the instructors in a constructive and cooperative environment designed to foster deeper understanding of change in the environment. All worked together to think theoretically and to support theory with data. These courses were an opportunity to learn the sorts of advantages and difficulties there are in working, at the same time, with global databases, spreadsheets, and geographic information systems.

We thank the students in these courses:

the experimental course (SNR&E 501), Fall, 1992.

Dawn M. Anderson, Katharine A. Duderstadt, Eugene A. Fosnight, Katharine Hornbarger, Deepak Khatry, Catherine MacFarlane, Gary Stahl, Stephen Uche, Hurng-jyuhn Wang, and Ruben De la Sierra.

the first year of the regular course (NRE, SPH 545), Fall, 1993.

Tatiana Bailey, Sanjay S. Baliga, Brent C. Blair, Tamara Carnovsky, John Castanon, Juan Carlos Cervantes, Bruce Frayne, Ilia E. Hartasanchez H., Roy Rojas Montero, Kameshwari Pothukuchi, Rhonda Ryznar, Suzy Salib, Caroline Stem, Kim Stone, Amy D. Sullivan, and Noreen White.

Others have offered substantial support and guidance in less tangible, but no less important, ways.

Kenneth H. Rosen of A.T.&T. Bell Laboratories first encouraged Sandra Arlinghaus to consider the idea of writing handbooks and put her in touch with Wayne Yuhasz at CRC Press. How suitable it is that the practical experience that Ken and Sandy shared as Lecturers of Mathematics at The Ohio State University should eventually turn into a practical handbook!

The generosity of Donald F. and Alma S. Lach of Chicago, parents of S. Arlinghaus, has made it possible to produce this, and subsequent, handbooks. We thank them for their continuing interest and wisdom, with special thanks from the Editor-in-Chief!

Frederick L. Goodman of the School of Education of The University of Michigan has been an enthusiastic supporter of the broader idea of education and maps. We thank him for his interest and input.

Community Systems Foundation (CSF) of Ann Arbor has generously provided access to both human and software resources for this project. We thank CSF, for its outstanding effort.

Digicopy Corporation donated professional copyright procurement services; we thank William E. Arlinghaus, President and CEO of Digicopy, for his customary flawless service.

The electronic databases of the World Bank and the World Resources Institute were an invaluable asset; we thank all those who have made these marvelous tools available to the public.

The American Geographical Society continues its generous policy of allowing authors to reprint from their own articles without charge (but of course with permission); we thank Douglas R. McManis for his careful attention to archival matters. We also thank the editorial staff of *Economic Geography*, in the Department of Geography at Clark University, for their permission to use the classic railroad maps of Mark Jefferson.

During the course of the past two years, during which this physical document was written, there are many with whom we have had conversations. To those friends, and colleagues who have written a brief e-mail note of encouragement, we offer our thanks. In particular, we thank Nanette R. LaCross and Gwen L. Nystuen, for their patience.

Finally, Wayne Yuhasz, Executive Editor--Physical Sciences at CRC Press, has made all of this possible. Wayne is delightful to work with; his good ideas, patience, and broad understanding of the sciences have made the writing of this book a pleasure.

TABLE OF CONTENTS

Simple cubic spline curve fitting
Interpolation using a cubic spline
Feigenbaum's graphical analysis

INTRODUCTION

THEORETICAL BACKGROUND

Overview

Computers offer even the casual user opportunities to handle data that were seldom dreamed of a decade ago. Most spreadsheets and other software packages that have some sort of analytic capability offer an option to graph input data; indeed, even many hand-held calculators do so. Default graphs generated by the computer often link points in a kind of follow the dots fashion; while this is a useful feature, default curves of this sort do not generally assign an equation to the curve. Thus, these curves cannot be projected nor can they be used for systematic interpolation between values. Many of the same software packages that offer default curves also permit the user to insert equations and to generate other graphs based on these equations. The problem is that the user new to this world often does not have the needed mathematical background. We offer a practical approach to showing how to fit curves; it is practical because it uses global real-world data and exposes the reader, as an important side benefit, to the problems encountered when using first-rate electronic (or other) data bases. The reader who is content to fit curves without some sort of mathematical discussion behind those efforts, but who understands that different choices of curves can forecast highly diverse alternative futures, should be able to use all the techniques in this handbook. This section is included because it is important that it be here should one wish it.

With any data set (presented in electronic or paper format), it is important first to examine the set for interesting or unusual patterns in the display. These patterns often influence decisions in choosing subsets of data and tools to analyze subsets. Thus, we encourage readers to approach any data set with a set of basic guidelines in mind and to browse it in a thoughtful manner. We offer the following as one set of guidelines; it will be repeated throughout, with commentary relating it to the databases selected in particular chapters.

PATTERNS IN DATA--WHAT TO LOOK FOR

1. What is the general organizational scheme of the entire set? Is it arranged alphabetically, numerically, or in some other fashion?

2. Are the real-world entries in the Table (nations, states, counties) expressed as comparable units? For example, county data and national data are generally not comparable.

3. Are the numerical entries in the Table expressed in comparable units? For example, data in one column might measure percentages while data in another column might measure thousands of dollars--these columns would not be comparable.

4. Are there gaps in the data? If so, what is their significance to the questions you wish to have the data answer?

Fitting a straight line

BLACK BOX SUMMARY
see Introduction for theoretical explanation

LEAST SQUARES REGRESSION LINE

$$y=mx+b$$

where
m is the slope of the line, or the "x-coefficient"
b is the y-intercept, or the "constant."

This Black Box Summary will appear in the text whenever this technique is employed. What we offer here is some background as to how it works. In this case, we motivate the theory behind it with an example.

Given the sample set of points, (1,1), (2,3) and (3,3). Fit a line y=mx+b to this set of points (m is the slope of the line, b is the y-intercept).

Thus, the following points are on the line y=mx+b:

(1, m+b)
(2, 2m+b)
(3, 3m+b).

The displacements of the sample set of points, from the points on the line, are the vertical distances between the points:

$$d[(1,1), (1, m+b)]=|m+b-1|$$
$$d[(2,3),(2,2m+b)]=|2m+b-3|$$
$$d[(3,3),(3,3m+b)]=|3m+b-3|.$$

A line of best fit might minimize these distances. So that dealing with absolute values does not become difficult, square the distances and then minimize the squares (hence, "least squares"). Note that true distance, rather than vertical distance, might have been used; it too is more difficult to handle, as would be "great circle," or some non-Euclidean form of distance.

What is now required is to minimize a function of two variables.

Minimize:
$$f(m,b)=(m+b-1)^2+(2m+b-3)^2+(3m+b-3)^2.$$

Simplifying:
$$f(m,b)=14m^2+12mb-32m+3b^2-14b+19.$$

Take the partial derivative of this last equation, with respect to m and with respect to b--cutting the surface with tangent planes in two directions to obtain two different equations in m and b. The strategy of cross-cutting to gain extra information is a good one and it is one that will be employed in a variety of contexts, other than the mathematical, throughout this book.

Continuing:
$$Partial[df/dm] = 28m+12b-32$$
$$Partial[df/db] = 12m+6b-14.$$

Since we are looking for a minimum, and since the partial derivative measures the slope of the plane tangent to a surface at a point, set the partial derivatives equal to 0 and look at values on either side to determine if there is a maximum (as there would be at the vertex on a concave down paraboloid), a minimum (as there would be at the vertex on a concave up paraboloid), or a saddle point (as there would be on a hyperbolic paraboloid).

Set:
$$Partial[df/dm\}=Partial[df/db]=0.$$
The values on either side are such that there is a minimum, as desired.

Solve:
$$28m+12b-32=0$$
$$12m+6b-14=0.$$
Therefore, m=1, b=1/3 and so
$$y=x+1/3$$
is the least squares best fit.

Spreadsheets have the capability to produce least squares output without understanding what is happening; however, they are fast and very useful. We shall make considerable use of this feature of Lotus 1-2-3, release 2.3 throughout the book and offer detailed information on how to use it. After one enters data in a spreadsheet and runs the regression/least squares feature, usually on-line help is sufficient to lead the beginner through the mechanics, an array of information is produced that may be called "an analysis of variance" or "regression output" or some similar phrase. What it gives generally is some or all of the following information.

the y-intercept (constant), to be used as the b-value in the least squares fit.

the x-coefficient, to be used as the m-value in the least squares fit.

the r-squared value; r is the coefficient of correlation--perfect correlation obtains a score of 1; no association, a score of 0. The r-squared value is the coefficient of determination and it too measures the degree of association from 0 to 1--it approaches 1 more slowly than does r, so that there is a wider spread in r-squared and distinctions are more readily made. Generally, a tight linear scatter of dots has an r-squared close to 1. Check the calculated r-squared against visual impression of the scatter of dots. Generally, these software packages only calculate correlation for ordinal data; they do not calculate non-parametric statistics on ranked, non-ordinal data (as would Spearman's correlation coefficient).

the standard error of the y-estimate--the amount of random error that might occur in the estimation of the y-intercept. The standard error may be used to draw a buffer of width one standard y-error parallel to and on either side of the regression line. Generally, about 68% of the time a y-estimate will fall within this band. If another buffer is created around the regression line at a distance of 2 standard y-errors, about 95% of the time a y-estimate will fall within this band. Roughly, the standard y-error is to the regression line as standard deviation is to the mean.

the standard error of the x-estimate--the amount the slope of the least squares line might be off due to random error.

Exponential curve-fitting

BLACK BOX SUMMARY
see Introduction for theoretical explanation

EXPONENTIAL CURVE
$$y = Ce^{ax} + b$$
where
$$a < 0$$
and
$y=b$ is the lower bound of the exponential;
C is a constant.

The idea behind exponential curve fitting is not hard once one understands how to use least squares and has the following definition.

Basic Definition:
$$\log_a y = x \text{ if and only if } a^x = y.$$

This definition is critical; it permits conversion of logarithmic to exponential functions and vice-versa.

There are a number of natural quantities, ranging from population to the amount of radioactivity, to interest compounded on a bank account, whose rate of growth or decay at any time is viewed to be proportional to the amount of that quantity present. This is the assumption that underlies unbounded population (or other) growth.

Assumption: The rate of population growth or decay at any time t is proportional to the size of the population at t.

Suppose Y(t) represents the size of a population at time t. According to the Assumption, the rate of growth of Y(t) is proportional to Y(t). Stated formally:
$$dY(t)/dt = kY(t)$$
where k is a constant of proportionality.
Separate the variables to solve this differential equation:
$$dY(t)/Y(t) = k \, dt.$$
Integrating:

$$\int 1/Y(t) \ dY(t) = \int k \ dt.$$

Therefore,

$$\ln |Y(t)| = kt + c(0)$$

where $c(0)$ is a constant of integration arising from the indefinite integration. Consider only the positive part, so that

$$Y(t) = e^{kt+c(0)}.$$

Let $Y(t_0) = e^{c(0)}$. Therefore,

$$Y(t) = Y(t_0)e^{kt};$$

exponential growth is unbounded as t approaches infinity. At the other extreme, suppose t=0. Thus,

$$Y(t) = Y(t_0)e^0 = Y(t_0).$$

Thus, $Y(t_0)$ is the size of the population at t=0, under conditions of growth where k>0--$Y(t_0)$ is the starter set. When k<0, the equation above describes decay.

Application of exponential growth--Law of Malthus, the rule of 70
 Under an assumption of exponential growth when will a given population double? $Y(t_0)$ is the starter set. When will the left side of the exponential growth equation be twice that? Or, when will it be the case that:

$$2Y(t_0) = Y(t_0)e^{kt} \ ?$$

Solving:

$$2 = e^{kt}$$

and using the basic definition yields:

$$\ln 2 = \ln e^{kt}$$

or

$$\ln 2 = kt.$$

Thus,

$$t = (\ln 2)/k = 0.6931472/k.$$

The value k, the constant of proportionality, is a decimal; multiply numerator and denominator by 100. The denominator is percent growth per year and the numerator is about 70--hence "the rule of 70" for making quick calculations about population doubling time.

 What is a corresponding rule for population tripling? Again:

$$t = (\ln 3)/k = 1.0986123/k * 100/100.$$

The population tripling time, under assumptions of exponential growth, is about 110 years. This strategy generalizes easily.

Logistic curve fitting

BLACK BOX SUMMARY
see Introduction for theoretical explanation

LOGISTIC CURVE

$$y = q/(1 + ae^{bx})$$

where

$$b < 0$$

and

$y=q$ is the upper bound of the curve, chosen ahead of time;
a and b are constants, calculated from the data.

In exponential growth, the underlying assumption is that the rate of population growth at any time t is proportional to the size of the population at t. Exponential growth is unbounded; in contrast, logistic growth is bounded. In logistic growth, the exponential assumption is modified to include the idea that in reality, when the population gets large, environmental factors dampen growth.

The growth rate decreases--$dY(t)/dt$ decreases. So, assume that the population size is limited to some maximum, q, where $0 < Y(t) < q$. As $Y(t)$ approaches q, it follows that $dY(t)/dt$ approaches 0 so that population size tends to be stable as t approaches infinity. The model is exponential in shape initially and includes effects of environmental resistance in larger populations. One algebraic expression of this idea is

$$dY(t)/dt = kY(t)\,(q - Y(t))/q$$

because in the factor $(q-Y(t))$, when $Y(t)$ is small, $(q-Y(t))/q$ is close to 1 (and the growth therefore close to the exponential) and when $Y(t)$ is close to q, $(q-Y(t))/q$ is close to 0, and the growth rate $dY(t)/dt$ tapers off. This factor acts as a damper to exponential growth.

Replace k/q by K so that

$$dY(t)/dt = KY(t)(q - (Y(t))$$

and the rate of growth is proportional to the product of the population size and the difference between the maximum size, q, and the population size.

Solve this last differential equation for $Y(t)$, separating the variables:

$$dY(t)/(Y(t)(q - Y(t))) = K\,dt.$$

Integrating:

$$\int dY(t)/(Y(t)(q-Y(t))) = \int K \ dt \ .$$

Use a Table of Integrals on the rational form in the left-hand integral:
$$1/q \ \ln|Y(t)/(q-Y(t))| = Kt + C$$
$$\ln|Y(t)/(q-Y(t))| = qKt + qC.$$
Because $Y(t) > 0$ and $q-Y(t) > 0$,
$$\ln \ (Y(t)/(q-Y(t))) = qKt + qC.$$
Therefore,
$$Y(t)/(q-Y(t)) = e^{qKt+qC} = e^{qKt}e^{qC} \ .$$
Replace e^{qC} by A. Therefore,
$$Y(t)/(q-Y(t)) = Ae^{qKt} \ ;$$
$$Y(t) = (q-Y(t))Ae^{qKt} \ ;$$
$$Y(t) = qAe^{qKt} - Y(t)Ae^{qKt} \ ;$$
$$Y(t)(Ae^{qKt} + 1) = qAe^{qKt} \ ;$$
$$Y(t) = qAe^{qKt} \ / \ (Ae^{qKt} + 1) \ ;$$
now divide top and bottom by Ae^{qKt} which is equivalent to multiplying the fraction by 1, so that
$$Y(t) = q/(1+(1/Ae^{qKt})) = q/(1+(1/A)e^{-qKt}) \ .$$
Replace $1/A$ by a and $-qK$ by b producing a common form for the logistic function
$$Y(t) = q/(1+ae^{bt})$$
with $b < 0$ because $b = -qK$, and $K > 0$.

There are a number of useful facts concerning the logistic equation.

The line $Y(t) = q$ is a horizontal asymptote for the graph.
    This is so because, for $b < 0$, the limit as t approaches infinity of $q/(1+ae^{bt})$ approaches $q/(1+a(0))$ which is q.
Can the curve cross this horizontal asymptote? Or, can it be that
$$Y(t) = Y(t)/(1+ae^{bt}) \ ?$$
Or, can it be that
$$1 = 1 + ae^{bt} \ ?$$
Or, that
$$ae^{bt} = 0?$$
Or, that $a = 0$? No, because $a=1/A$. Or, that $e^{bt}=0$--no, this cannot happen.
    Thus, the logistic growth curve described above cannot cross the horizontal asymptote. The logistic curve approaches the horizontal

asymptote entirely from one side; there is no possibility of overshooting it and coming back to settle down on the other side, nor is there the possibility of overshooting, and oscillating back and forth across the asymptote with continually smaller overshooting of the asymptote.

It is also useful to know the coordinates of the inflection point of the logistic curve.   The slope of the curve is steepest at the inflection point--the point where the curve changes from a concave-up exponential type of curve to a damped concave-down curve.

First, find the vertical component of the coordinates for the inflection point.   The equation $dY(t)/dt = KY(t)(q-Y(t))=KqY(t)-K(Y(t))^2$ is a measure of population growth. Find the maximum rate of growth--it is the derivative of the previous equation:

$$d^2Y(t)/dt^2 = Kq - 2KY(t).$$

To find a maximum (minimum), set this last equation equal to zero:

$$Kq - 2KY(t) = 0.$$

Therefore, $Y(t) = q/2$--the inflection point is halfway up, between the x-axis and the horizontal asymptote at y=q. This is the vertical coordinate of the inflection point of the curve for $Y(t)$, the logistic curve--one might therefore choose q as some sort of doubled time from the present, when the present is at the x-value of the inflection point. Since $dY(t)/dt$ is increasing to the left of q/2 ($d^2Y(t)/dt^2 >0$) and $dY(t)/dt$ is decreasing to the right of q/2 ($d^2Y(t)/dt^2 < 0$), the maximum rate of growth occurs at $Y(t) = q/2$. (The rate at which the rate of growth is changing is a constant since the first differential equation is a quadratic (its graph is a parabola).)

Next, find the horizontal component of the inflection point. To find t, put $Y(t) = q/2$ in the logistic equation and solve:

$$q/2 = q/(1+ae^{bt}) .$$

Solving,

$$1+ae^{bt} = 2$$
$$e^{bt} = 1/a$$
$$-bt = \ln a$$
$$t = (\ln a)/(-b).$$

Thus, the coordinates of the inflection point of the logistic curve are:

$$((\ln a)/(-b), q/2).$$

Simple cubic spline curve fitting

BLACK BOX SUMMARY
SIMPLE CUBIC SPLINE CURVE FITTING
Fit a cubic, $S(x)$, of the form
to each interval
between a finite set of given evenly-spaced
sample points, one unit apart,
$(x_1,y_1),...,(x_n,y_n)$.
The spline, $S(x)$, composed of the following n equations

$$a_1(x-x_1)^3+b_1(x-x_1)^2+c_1(x-x_1)+d_1$$
$$a_2(x-x_2)^3+b_2(x-x_2)^2+c_2(x-x_2)+d_2$$

$$........$$

$$a_{n-1}(x-x_{n-1})^3+b_{n-1}(x-x_{n-1})^2+c_{n-1}(x-x_{n-1})+d_{n-1}$$

that will fit the sample points has coefficients given by:
$$a_i=(M_{i+1} - M_i)/6$$
$$b_i=M_i/2$$
$$c_i=(y_{i+1} - y_i) - ((M_{i+1} +2M_i)/6)$$
$$d_i=y_i$$
where $i=1,2,...,n-1$
and where the M_i are determined as solutions
to the following matrix equation

$$
\begin{bmatrix}
1 & 0 & 0 & 0 & \dots & 0 & 0 & 0 \\
1 & 4 & 1 & 0 & \dots & 0 & 0 & 0 \\
0 & 1 & 4 & 1 & \dots & 0 & 0 & 0 \\
\multicolumn{8}{c}{\dots\dots} \\
0 & 0 & 0 & 0 & \dots & 1 & 4 & 1 \\
0 & 0 & 0 & 0 & \dots & 0 & 0 & 1
\end{bmatrix}
\begin{bmatrix}
M_1 \\ M_2 \\ M_3 \\ \dots \\ M_{n-1} \\ M_n
\end{bmatrix}
= 6
\begin{bmatrix}
0 \\ y_1 - 2y_2 + y_3 \\ y_2 - 2y_3 + y_4 \\ \dots \\ y_{n-2} - 2y_{n-1} + y_n \\ 0
\end{bmatrix}
$$

where the Ms are a column matrix (vector) that is multiplied on the left
by an (nxn) matrix of constants; this product is equal to 6 times a
column matrix of constants whose values can be determined from the
second coordinates of the given sample points. Thus, to solve for the
column of Ms, it is necessary to find the inverse of the (nxn) matrix of
constants on the far left.

The cartographer's spline can be imitated using mathematics: hence the name "spline." When the mathematical spline is composed of pieces of polynomials of degree three, cubics, the procedure is cubic spline fitting. Roughly, the idea is to fit pieces of cubic curves between a finite set of sample data points: one cubic is fit between two adjacent sample points and another cubic is fit between another pair of adjacent sample points. In this manner, one can continue piecing together a curve between a finite set of pairs of adjacent points. Because the curve is fit between points, and never extended beyond any sample point, it is a fit that is bounded and is useful only for interpolation--not for extrapolation.

The fit should be such that the curve is smooth and continuous throughout its bounded interval, and so that at the sample points, where the curve is spliced together, a line tangent to the curve has the same slope whether the line is tangent to the curve determined by a cubic to the left of the sample point or whether the line is tangent to the curve determined by a cubic to the right of the sample point. The curve is differentiable.

The Black Box Summary above gives the highlights of cubic spline curve fitting; readers interested in understanding the detail of why these equations work are referred to various texts, to handbooks with computer programs for the myriad variations on splines, and to scholarly articles on splines.

References

1. Birkhoff, G., De Boor, C. R., de Boor, C. R. Piecewise polynomial interpolation and approximation. *Approximation of Functions (Proc. Sympos. General Motors Res. Lab.*, 1964) 164-190. Elsevier Publ. Co., Amsterdam, 1965.
2. de Boor, C. R. The condition of the B-spline basis for polynomials. *SIAM Journal on Numerical Analysis,* 25, 1, 148-152, 1988.
3. Press, W. et al. *Numerical Recipes in C: The Art of Scientific Computing*, Cambridge, New York, 1988.
4. Sprott, J. C. *Numerical recipes: routines and examples in BASIC*, Cambridge, New York, 1991.

CHAPTER 1

POPULATION DATA ANALYSIS

ANALYTICAL TECHNIQUES/TOOLS USED:

Straight line curve-fitting--least squares
Exponential curve-fitting
Exponential curve-fitting with a lower bound
Logistic curve-fitting
Gompertz curve-fitting

DATA TYPE: ABUNDANT LONGITUDINAL DATA

transfer of data from data base to spreadsheet
cleaning, analysis, and graphing of transferred data

Overview of Data

Population data sets were chosen as the initial sets of data on which to illustrate curve fitting because there are so many data sets concerning population and because there is a vast systematic literature already available concerning the analysis of these data. Indeed, demography is the science of the statistical analysis of human populations, focusing on size and density, distribution, and vital statistics. It is a long-standing science, certainly dating from the time of Malthus; Malthus's basic law states that in the absence of constraints on resources or other factors, one would naturally expect exponential population growth. A larger population generates yet a larger one, in much the way that the compounding of interest causes a bank account to grow exponentially.

The science of population dynamics includes the study of births, deaths, and population age cohort changes; in and out-migration among countries; and, rural/urban migration within countries. A particularly exciting approach, cast in the current context of analyzing the human dimensions of global change, is to consider the relationship between population dynamics and environmental degradation. Some of the topics within the idea of environmental degradation include adverse toxic environmental impact; consumption of non-renewable resources (such as oil) at rates that exceed the capacity to replace them with an alternative in the future; and, the consumption of renewable resources at rates in excess of replacement--non-sustainable consumption. A variety of other components contribute to environmental degradation, including those coming from economic and other development; in some cases populations dynamics is the dominant factor, in others, development is the driving force behind environmental degradation. The data sets chosen in this book are typical of the sorts of data sets that might contribute significantly to analyzing the relationship between population dynamics and environmental degradation. Thus, the reader who is learning to fit curves, in any disciplinary context, also has the opportunity to see how data might be used creatively in the emerging science of population dynamics.

Various international organizations publish data on population in an electronic format. There is no global standard for census taking: the collection of data concerning human populations. Thus, there will be variation in censuses on numerous bases: from the national level, to the state level, to the county level.

With any data set (presented in electronic or paper format), it is important first to examine the set for interesting or unusual patterns in the display. These patterns often influence decisions in choosing subsets of data and tools to analyze subsets.

PATTERNS IN DATA--WHAT TO LOOK FOR

1. What is the general organizational scheme of the entire set? Is it arranged alphabetically, numerically, or in some other fashion?

2. Are the real-world entries in the Table (nations, states, counties) expressed as comparable units? For example, county data and national data are generally not comparable.

3. Are the numerical entries in the Table expressed in comparable units? For example, data in one column might measure percentages while data in another column might measure thousands of dollars--these columns would not be comparable.

4. Are there gaps in the data? If so, what is their significance to the questions you wish to have the data answer?

The World Resources Institute data base contains information, derived from World Bank data, for most countries of the world. This information is partitioned on the basis of over 500 variables, many of which are related to population. There is longitudinal data available, in electronic format, for many of the population indicators. With this electronic data base, as with others, the content of interest can be selected and transferred to commercial packages of choice. This feature makes these databases extremely useful for analysis; occasionally, computer users have some difficulty implementing these data bases. Thus, to begin, we illustrate in detail (by example), how to transfer information from the World Resources Institute data base into a Lotus 1-2-3 spreadsheet. Then, we suggest how to analyze data within a spreadsheet, and finally, how to portray some elements of the data set on a map. The general strategies function well with other software although they are, in terms of detail, unique to the software selected.

Data base from the World Resources Institute

A traditional idea in studying population is that an increase in population will result in an increase in births. Suppose that one wished to investigate this theme, of the relationship between total population and total births, from a systematic standpoint. One logical beginning might be with a major international data base.

Two of the over 500 variables in the World Resources Institute data base, derived from public domain information, concern total population by country, expressed in units of thousands, and crude birth rate per thousand population by country. For the sake of example, we chose to look at these two variables for the country of Bangladesh (formerly East Pakistan)--a country known to have an interesting history in terms of population growth.

Transfer of information from data base to spreadsheet

Most electronic databases are very easy to use--the online help guides even the shyest computer user through the mechanics of selecting precisely the desired information for the desired countries. If difficulties in using these databases arises, it is usually, as with any software, in the seam of the interface linking one piece of software to another.

Remember, depending on the operating system in use, that it is often useful to run a directory on the software to see what is available, prior to starting an analysis of data. In a PC environment, the command, at the C prompt of the subdirectory holding the software, of dir/w, will cause the content of the directory for the software to be displayed in table-format across the width of the screen. The command dir/p will cause the content of the directory for the software to be displayed in column-format, accompanied by the size of the file and the date of last use, on the screen a page at a time. It is easy to forget how to start databases--especially in environments loaded with different software, some of which the user employs only occasionally. Looking up names in a directory removes the need to remember commands specific to individual software packages and replaces it, instead, with the universal idea of looking it up. Directories are powerful tools.

A set of data chosen to examine some theme is a start. The next step is to prepare to analyze the data. Many databases have some analytical capabilities built-in. They are often easy to use but are currently likely not to have the power of the commercial database programs that have been refined for years to present the user with sophisticated analysis and an online help facility that is easy to use (in any package, try F1 if it is not clear how to gain access to the online help). It may also be the case that graphics displayed in databases on the screen may not print out directly on a printer. In brief, there may be a number of shortcomings with the analysis segment of a data base; the primary function of the data base is as a convenient source of data, often expressed at the scale of individual nations or groupings of nations, concerning a number of different variables.

To transfer data from an electronic data base to a spreadsheet, it is necessary to "export" the data from the database to the spreadsheet. The data base should have online help that is easy to use to create a file to export. If there is any difficulty, it is likely to come in knowing how to name the file so that it appears in the target spreadsheet software when that is next opened up. In the World Resources Institute data base, the following strategy works well. When the time comes to name the file in the process of exporting, the user is offered a line of text including c:\wrd\........ . The dots indicate that the user is to supply a file name. If the file name is simply appended, the file will not go directly to the spreadsheet. Suppose the spreadsheet software is Lotus 1-2-3, release 2.3, stored in subdirectory 123r23. Suppose the name of the spreadsheet is to be "sample-1.wk1--the extension "wk1" is specific to the spreadsheet software. Now, when the database software presents the line c:\wrd\......., use the backspace to erase "wrd\" and substitute 123r23\sample-1. Now the line of text will read c:\123r23\sample-1. When the type of file this is to be stored as is selected as a "wk1" file, the data in "sample-1" will now appear as one spreadsheet that can be selected directly from Lotus 1-2-3, release 2.3.

We have chosen to use this particular spreadsheet with the World Resources Institute data base. Choose your source of data first; then determine which other analytical tools it can interact with. Then try the various interfaces, determine which will work best for you, and proceed with analysis. Minimal facility with various spreadsheets and analytical software is important in making a good selection, just as exposure to various natural, artificial, and formal languages can foster a broad, general understanding of mathematical and natural language.

Cleaning of data transferred from an electronic database

The four steps enumerated in the box above may seem a bit obvious but are very important in making sure that results are as clear and as accurate as possible. When dealing with data that has been difficult to collect, it is important not to introduce any errors in the analysis. One reason direct transferral of data from electronic data base to spreadsheet (or other software) is important is that no re-keying of the data is involved and therefore no new errors can be introduced from any extra data entry. However, when different variables, or different countries are chosen from the electronic data base, the entries may have to be rearranged or manipulated in some other way so that desired analyses and comparisons can be performed.

To illustrate the idea of "cleaning" data, suggested in steps 1-4 above, we chose a very simple example--one country and two variables. With more geographical units and more variables, the increase in difficulty of cleaning is more likely exponential than linear! The necessity to clean data is of critical and fundamental importance.

1. What is the general organizational scheme of the entire set? Is it arranged alphabetically, numerically, or in some other fashion?

The entire set of World Resources Institute Data can be sorted first by variable and then by country, or first by country and then by variable. The capability to sort the data in both ways makes this database one that is quite flexible and one that can be used to interpret a variety of themes.

2. Are the real-world entries in the Table (nations, states, counties) expressed as comparable units? For example, county data and national data are generally not comparable.

This data base contains data for most countries of the world; some variables have data for more countries than do others. Thus, when looking at more than one variable at a national level, one must be certain that exactly the same set of countries is being dealt with, prior to making comparisons. There is also data grouped at regional levels--that is as groupings of countries, such as "oil producing exporters." Such groupings are nice to have, but they can introduce bias by suggesting what to study.

TABLE 1.1

Bangladesh; actual data, 1955-1990, projected data, 1995-2025.
Total population and crude births in millions
Source: World Resources Institute Data Base.

STRAIGHT LINE FIT

Actual data, 1955–1990   Bangladesh

Year	Pop. mil.	Crude births, mil
1955	45.486	2.137842
1960	51.419	2.406409
1965	58.312	2.723170
1970	66.671	3.166872
1975	76.582	3.714227
1980	88.219	4.163936
1985	101.147	4.531385
1990	115.593	4.878024
1995	132.219	5.368091
2000	150.589	5.752499
2005	170.138	5.971843
2010	188.196	5.532962
2015	204.631	5.074848
2020	220.119	4.842618
2025	234.987	4.69974

3.   Are the numerical entries in the Table expressed in comparable units? For example, data in one column might measure percentages while data in another column might measure thousands of dollars--these columns would not be comparable.

The entries in this data base are expressed both as raw data and as rates. Thus, one must be careful not to compare a rate to raw data, or to average a set of rates (producing meaningless results). The variable concerning total population is expressed as raw data; the variable concerning crude births is expressed as a rate per 1000. To make the variables comparable, it is an easy matter simply to multiply each of the crude birth rates by the number of thousands of population per country in order to get a figure on total crude births. The data in Table 1.1 has been adjusted, decimally, so that the total population figures in the second column show millions of total population for the country of Bangladesh over a seventy year period, and so that the crude birth figures show millions of total crude births for Bangladesh for the same seventy year period.

4. Are there gaps in the data? If so, what is their significance to the questions you wish to have the data answer?

In Table 1.1, the first column shows years for which there is data. In fact, more was available. From the period for 1970 to 1990, data was presented on an annual basis. When fitting curves, it is generally appropriate to have evenly-spaced units on the x-axis represent evenly-spaced units in the data. To introduce even spacing in this data set, one might choose only to look at the data for which there is annual data, or one might choose only to look at data every five years. We chose the latter course.

Note also in the first column of Table 1.1, that the years from 1955 to 1990 are positioned differently than those from 1995 to 2000. This offset pattern is presented to remind the user of the Table that the data to 1990 is actual data, and that beyond 1990 is projected data.

Predictions of any sort are at best tenuous; when funding is allocated based on projections, questions should be asked. The more analytic support one has for asking questions, including charts, graphs, maps, and other visual devices, the more likely one is to achieve some sort of constructive, interactive communication.

Analysis of spreadsheet data

There are a variety of default analyses that can be performed easily using spreadsheets; bar charts can be drawn, pie charts can be created, and curves can be drawn. When curves are drawn, the default curve is simply a curve formed by joining discrete values with line segments--a follow-the-dots approach. Figure 1.1 shows default curves when the actual data, from 1955 to 1990, from Table 1.1 are graphed. Default graphs offer an easy and effective visual display. They do not have an equation(s) associated with them, so they cannot be referred to in notation (without additional work) and therefore they cannot be used to make projections from actual data. Consequently, their only use is as visual backup--when there are also equations behind the curves, then they serve as analytic tools, as well. Thus, it becomes important to be able to associate equations, which can then generate suitable curves, with actual data.

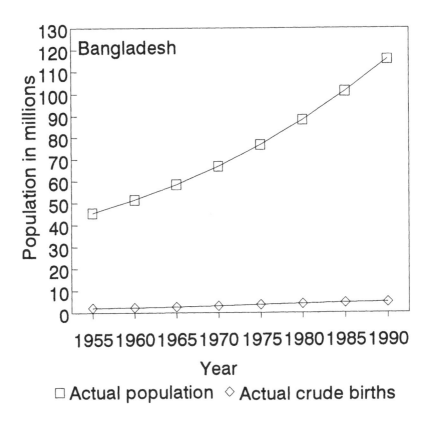

Figure 1.1. Total population and total crude births, Bangladesh, 1955-1990.

The graph in Figure 1.1 shows a curved rise in the population of Bangladesh over the time period from 1955 to 1990. One might therefore suppose that projections made on the basis of this actual data would continue to reflect this trend. The crude birth data also shows a slight, but steady, rise during this time period, too. The database from which these values were extracted also contains data projected to the year 2025 for these variables. When these are graphed, Figure 1.2, the population clearly continues to rise, although the rate at which it rises appears to taper off a bit. The birth rate data is more difficult to see at the scale of the graph in Figure 1.2. When the scale is enlarged, Figure 1.3, it becomes clear that a steadily rising trend of actual data is projected as one that rises for a short time in the near-future and then drops off sharply at the year 2005. These default graphs, alone, suggest the importance of questioning why there is an assumption made about a decline in births in Bangladesh starting in 2005. Is it the result of some current programs involving education and family planning that are already in place, or is it a wished-for result that could come about IF certain proposals involving education and family planning are funded? The difference is substantial in terms of policy implications; to understand which it is, it is critical to know on what basis this drop is forecast.

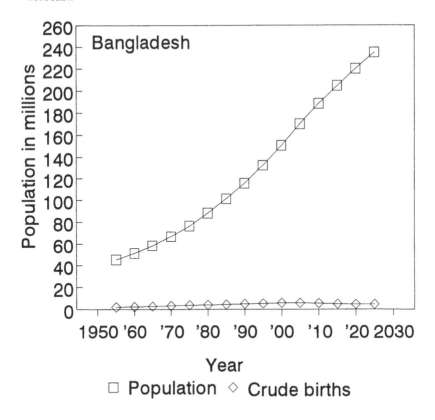

Figure 1.2. Total actual and projected population and crude births, 1955-2025.

The mere default graphing of data can suggest critical questions; graphing beyond the default level can provide even more insight. We consider a few additional ways to look at data using tools readily available in spreadsheets. First each variable is considered graphically and then the theme of the relationship between total population and births is returned to.

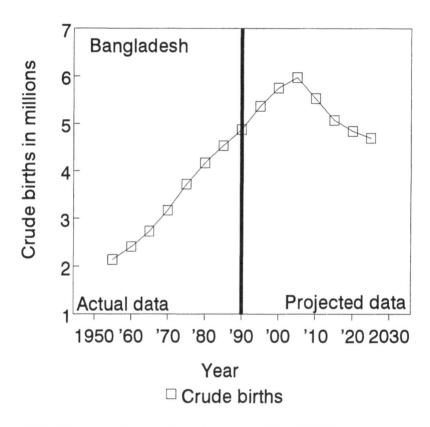

Figure 1.3. Bangladesh, actual and projected total crude births: 1955-2025.

Straight line curve fitting--least squares analysis--birth data

The graph of the variable, "total crude births," suggests that a straight line might fit the actual data quite well. In Table 1.2, a straight line has been fit to the actual data from 1955 to 1990, using least squares "regression" analysis. The fit is quite tight, as is indicated by the r-squared value of 0.99 (Table 1.2). If a different fit had been tried, resulting in a value of r-squared smaller than this, one might conclude that the curve with the value of r-squared closest to 1 gives the tightest fit to the scatter of dots. What that means would depend on the kind of data used as inputs; experimental data of unknown reliability produces r-squared values of questionable meaning, as does data that is based on averages.

The procedure for obtaining this line is straightforward and uses only the capability of a good spreadsheet. Select the regression feature from the spreadsheet. Enter the column, from 1955-1990, for "year" as the x-axis value; enter the crude births values for 1955-1990 as the next variable (the Y-range in Lotus 1-2-3, release 2.3). Choose a blank location of the spreadsheet as the output range. Then, go with the regression, and the output should be similar to that part of Table 1.2 below the entry for the year 2025. Typically, the last two lines, giving the phrase "linear fit" and the equation of the regression line are not part of the output. The user needs to understand the content of the output enough to know that generally, the equation of the line of least squares will be of the form

$$y=mx+b$$

where m is the slope of the line and b is its intercept on the y-axis. The "X Coefficient(s)" in Table 1.2, 0.0825614617, is the x coefficient--m-- in the displayed equation above. The "constant" in Table 1.2, -159.38724963, is the y-intercept and so is the b-value in the displayed equation.

As a specific sample, fit a straight line to the data in Table 1.2-- using Lotus 1-2-3 release 2.3, in which the year 1955 appears in column A, row 5 (cell A5) of the spreadsheet. A reader unfamiliar with using a spreadsheet might wish to actually try fitting a straight line by following the instructions below, step by step.

STRAIGHT LINE FIT TO THE DATA OF TABLE 1.2
(Refer to Table 1.2 and Figure 1.4)

1. Enter the years 1955-1990, for which there is ACTUAL data, in spreadsheet column (A) to be used as the x-axis input.

2. Enter the ACTUAL data that varies over time, 1955-1990, in another column (spreadsheet column B--total crude births in this case), to be used as y values

3. Choose the regression feature from the software, with the x values as in step 1 and the y values as in step 2.

4. Choose the output range as a blank area in the spreadsheet. Then proceed with the calculation as directed by the software; the output from the regression will appear in a form similar to the one in Table 1.2, bottom half (produced in Lotus 1-2-3, release. 2.3).

5. The equation below the output range must generally be derived by the user from the regression output. The slope-intercept form for the equation of a straight line (y=mx+b) is used. The "X Coefficient" from the regression output is used as "m". The "Constant" from the regression output is used as b. Thus, the equation, that fits the ACTUAL data from 1955-1990, can be read directly from the regression output.

6. In a separate column of the spreadsheet, labelled "projected straight line vals" in Table 1.2 (Column E), enter the equation derived from the regression: in cell E5, enter the formula 0.082561*A5-159.387. The value 2.019755 should appear at the top of the "projected straight line vals" column. Then, copy the cell content from E2 to the 7 cells below it; this should produce the numerical range, from 1955-1990, shown in the "projected straight line vals" column.

7. Graph the results; select an XY-graph. Put the 8 entries for the years in the X-range, and the eight actual values in the A range. Enter the 8 entries of the projected linear data, from 1955-1990, in an additional range, B, of the spreadsheet (Figure 1.4). The result should look like the left half of Figure 1.4.

Often, spreadsheets will present the user with more decimal places than desired. In that case, the user must decide whether to truncate the long decimal or round it up or down. Each strategy has merit; making a rational decision, employing it consistently throughout an analysis, and telling the reader what decision has been made and why it has been made is a good approach. In this example, decimals were truncated because it appeared consistent with what the computer does when the column is expanded; there is less chance of introducing an error by merely erasing digits--when old ones are altered the risk increases.

EXTRAPOLATION

To extrapolate the linear curves fit to the actual data, simply add data to spreadsheet columns and graph the extended columns. In the case here:

1. Extend the number of years to cover a period of time up to and including the year 2025 (in five-year intervals) as in Table 1.2, column A.

2. Enter WRD data for the entries from 1995 to 2025.

TABLE 1.2 (Source: World Resources Institute)
Least squares regression line fit to actual crude birth data, 1955-1990.
Right-hand column shows fit of line to actual data
and extrapolation to cover 1995-2025.

STRAIGHT LINE FIT		PROJECTED
Actual data, 1955–1990	Bangladesh	STRAIGHT
Year Crude births, mil		LINE VALS.
1955	2.137842	2.019755
1960	2.4064092	2.43256
1965	2.7231704	2.845365
1970	3.1668725	3.25817
1975	3.714227	3.670975
1980	4.1639368	4.08378
1985	4.5313856	4.496585
1990	4.8780246	4.90939
1995	5.3680914	5.322195
2000	5.7524998	5.735
2005	5.9718438	6.147805
2010	5.5329624	6.56061
2015	5.0748488	6.973415
2020	4.842618	7.38622
2025	4.69974	7.799025

Regression Output:

Constant	−159.38724963
Std Err of Y Est	0.089798274
R Squared	0.9932854447
No. of Observations	8
Degrees of Freedom	6
X Coefficient(s)	0.0825614617
Std Err of Coef.	0.0027712349

Linear fit:
$y = 0.082561 {}^{*}x - 159.387$

3. Copy the formula in cell E12 into cells E13 to E18. The Table should now be identical to Table 1.2.

4. Graph the extended data; simply open the previous graph and extend the columns for X, A, and B ranges to cover the values associated with the years 1995 to 2025 as well as those from 1955 to 1990. The resulting graph should appear as that in Figure 1.4. Alter labels to suit individual tastes.

The equation obtained from fitting a straight line to the actual data may then be used to calculate values for projected data. Table 1.2 shows projected values in the right hand column, obtained by entering the value 0.082561*x-159.387 into the first row of that column and replacing the value "x" with "1955" or with the cell address, A5, of the range containing "1955."   Subsequent values in this column were calculated by inserting appropriate values for the "Year" column into the formula derived from the regression line. Individual values for x may be inserted by hand; however, when the cell address is used instead, the formula may simply be copied from one range to another and new values are then calculated automatically--in Lotus 1-2-3, simply use the "copy" command from the main menu and copy the content of the top cell in the right-hand column to the remaining cells in that column. This latter strategy is easier and there is less room for the introduction of error.

When the data are graphed, it becomes clear that the linear fit is quite good for the actual data (Figure 1.4) and is good on the projected data until about the year 2000. After 2005, there is a sharp drop away from the least squares line.   The straight line fit offers concrete, analytical evidence on which to question this drop: "why does the projection veer off from the straight line which naturally fits this data?"

The sorts of follow-up questions that might ensue could include questioning where WRD data comes from; if it comes directly from the Bangladesh government, one might then try to determine the basis on which they make projections and the sorts of education and family planning projects that are already in place. Is this projected, radical downswing a result of carefully orchestrated planning that can be documented, or is it based on wishful thinking about what must take place for success in caring for this segment of the world's population? Indeed, it might be prudent to fit a curve to this projected segment from 2005, although any sort of projection is a very risky, but often used, strategy.

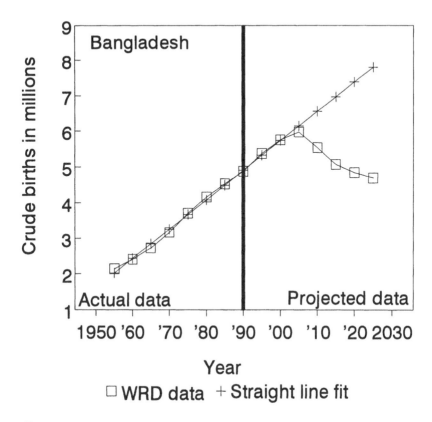

Figure 1.4. Least squares line--fit to actual and projected WRD data.

BLACK BOX SUMMARY
see Introduction for theoretical explanation

LEAST SQUARES REGRESSION LINE

y=mx+b

where
m is the slope of the line, or the "x-coefficient"
b is the y-intercept, or the "constant."

Exponential curve-fitting--crude birth data

The total crude birth curve in Figure 1.4, to the right of 2000 or 2005 is one that shows a sharp decline. To capture it in an equation, it seems natural to fit an exponential curve to it. The following strategy illustrates how to fit an exponential to a set of data; the reader might wish to follow along, step by step, using a computer and Lotus 1-2-3, release 2.3.

EXPONENTIAL CURVE FIT TO THE DATA OF TABLE 1.3
(Refer to Table 1.3 and Figure 1.5)

1. Enter the years in a spreadsheet column (A) to be used as the x-axis input. The first entry, for 1955, is in cell A5.
2. Enter the data that varies over time in another column (spreadsheet column B--WRD projection in this case), to be used as y values.
3. Take the natural log of the y values (other bases for the logarithms work, too). These appear in column C and are obtained from column B by writing @LN(B5) in cell C5. Copy the formula in C5 to cells C6 through C10.
4. Choose the regression feature from the software, with the x values as in step 1 and the y values as in step 3--the regression will be run on the x values and the LN y values (to obtain the exponential form).
5. Choose the output range as a blank area in the spreadsheet. Then proceed with the calculation as directed by the software; the output from the regression will appear in a form similar to the one in Table 1.3, bottom half.
6. The equations below the output range must generally be derived by the user from the regression output. The slope-intercept form for the equation of a straight line (LN y=mx+b) is used. The "X Coefficient" from the regression output is used as "m". The "Constant" from the regression output is used as b. Thus, the first equation can be read directly from the regression output.
7. The second equation, that is an exponential, is derived from the first equation by raising both sides to e, the base of the natural logarithms. This is the equation that will be used to enter projected exponential values, starting with the year 2000.

TABLE 1.3
Exponential curve fit to projected crude birth data, 2000-2025.
Horizontal asymptote at y=0.
(Source: World Resources Institute)

EXPONENTIAL FIT, BOUNDED BELOW AT Y=0

Projected data, 2000−2025

Year	WRD projection	ln col 2	exp. proj. 2000
2000	5.7524998	1.7496345082	5.9857885876
2005	5.9718438	1.7870557239	5.6977860779
2010	5.5329624	1.7107233683	5.4236406305
2015	5.0748488	1.6242967314	5.1626855215
2020	4.842618	1.5774554836	4.9142861058
2025	4.69974	1.547507188	4.6778382742

Regression Output:

Constant	21.513547691
Std Err of Y Est	0.0343425533
R Squared	0.9001961642
No. of Observations	6
Degrees of Freedom	4
X Coefficient(s)	−0.0098620798
Std Err of Coef.	0.0016418881

Linear equation:

$Ln(y) = -0.00986*x + 21.51354$

Exponential equation to fit 2000 to 2025:

$y = @exp(-0.00986*x + 21.51354)$

8. In a separate column of the spreadsheet, labelled "exp. proj. 2000" in Table 1.3 (Column D), enter the equation derived from the regression: in cell D5, enter the formula @EXP(-0.00986*A5+21.51354). The value 5.9857885876 should appear at the top of the "exp. proj. 2000" column. (The number of decimal places displayed will depend upon the width of the column.) Then, copy the cell content from D5 to the 5 cells below it; this should produce the entire numerical range shown in the D column.

9. Graph the results; select an XY-graph. Put the 6 entries for the years in the X-range, and the six WRD projected values in the A range. Enter the 6 entries of the projected exponential data in an additional range, B, of the spreadsheet (Figure 1.5). The result should look like Figure 1.5.

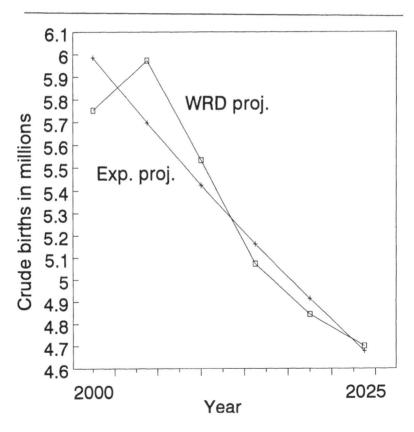

Figure 1.5. Graph of exponential fit to projected crude birth data, 2000-2025; the x-axis is a horizontal asymptote for the exponential curve.

TABLE 1.4
Exponential curve fit to projected crude birth data, 2005-2025.
Horizontal asymptote at y=0.
(Source: World Resources Institute)

EXPONENTIAL FIT, BOUNDED BELOW AT Y=0

Projected data, 2005–2025

Year	WRD projection	exp. proj. 2000	exp. proj. 2005
2005	5.9718438	5.6977860779	5.8819053674
2010	5.5329624	5.4236406305	5.5325246919
2015	5.0748488	5.1626855215	5.203896961
2020	4.842618	4.9142861058	4.8947894658
2025	4.69974	4.6778382742	4.6040427191

Regression Output:

Constant	26.327715447
Std Err of Y Est	0.0215939977
R Squared	0.9640366419
No. of Observations	5
Degrees of Freedom	3
X Coefficient(s)	−0.0122472991
Std Err of Coef.	0.0013657243

Linear equation:

$Ln(y) = -0.01224*x + 26.32771$

Exponential equation to fit 2005 to 2025:

$y = @exp(-0.01224*x + 26.32771)$

The exponential curve found in this case fits the WRD projections worst in the time interval from 2000 to 2005; thus, a logical next step to improve the fit might be to repeat the strategy using WRD data from 2005 to 2025. Table 1.4 shows the results of executing a regression on the years 2005 to 2025 and the associated natural logs of the WRD projections. These results are displayed in column D of the spreadsheet and are labelled "exp. proj. 2005" -- in contrast to the projected exponential starting in 2000. The spreadsheet in Table 1.4 shows the WRD projections, the exponential projections beginning in 2000, and the exponential projections beginning in 2005. The regression analysis calculated an R-squared value of 0.96 using the data beginning in 2005. This R-squared value is closer to 1 than is that of 0.90 derived from the regression beginning in 2000. The 2005 fit is tighter than is the 2000 fit.

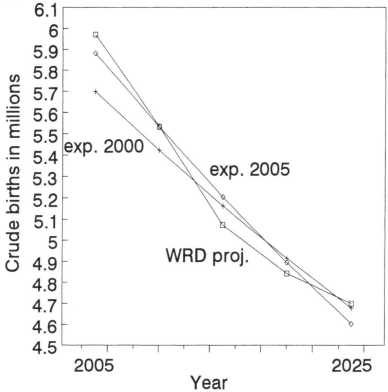

Figure 1.6. Graphs of exponential fits to projected crude birth data, 2000-2025, and 2005-2025; the x-axis is a horizontal asymptote for each exponential curve.

When all three curves are plotted (Figure 1.6), it also appears that the 2005 exponential projection offers a better fit to the WRD projection than does the 2000 exponential projection. In addition, the visual display suggests that better exponential fits than either of these might be available--it might be possible to adjust the R-squared value upward.

The exponentials above were fit to the data assuming the x-axis as a lower bound--as x approaches infinity, the curve will approach the value y=0. An exponential fit assuming y=0 as a lower bound is the simplest sort of exponential fit. An exponential curve can be fit given any real number, n, with y=n as a lower bound.

In an effort to make the fitted curves track more accurately the WRD projections, raise the lower bound, to y=4 (suggested by the data). The idea is to force the exponentials to begin tapering off toward the lower bound (horizontal asymptote). The summary below, when used in conjunction with Table 1.5, shows how to achieve this style of fit using y=4, instead of y=0, as a lower bound.

EXPONENTIAL CURVE FIT TO THE DATA OF TABLE 1.5
(Refer to Table 1.5 and Figure 1.7)

1. Enter the years in spreadsheet column (A) to be used as the x-axis input. The first entry, for 2000, is in cell A5.

2. Enter the data that varies over time in another column (spreadsheet column B--WRD projection in this case), to be used as y values.

3. Take the natural log of -4 plus the y values (other bases for the logarithms work, too). These appear in column C and are obtained from column B by writing @LN(-4+B5) in cell C5. Copy the formula in C5 to cells C6 through C10. The reason the value of -4 is introduced is because of the lower bound, y=4.

4. Choose the regression feature from the software, with the x values as in step 1 and the y values as in step 3--the regression will be run on the x values and the LN (-4+y) values (to obtain the exponential form).

5. Choose the output range as a blank area in the spreadsheet. Then proceed with the calculation as directed by the software; the output from the regression will appear in a form similar to the one in Table 1.5, bottom half.

6. The equations below the output range must generally be derived by the user from the regression output. The slope-intercept form for the equation of a straight line (LN (-4+y)=mx+b) is used. The "X Coefficient" from the regression output is used as "m". The "Constant" from the regression output is used as b. Thus, the first equation can be read directly from the regression output.

TABLE 1.5
Exponential curve fit to projected crude birth data, 2000-2025.
Horizontal asymptote at y=4.
(Source: World Resources Institute)

EXPONENTIAL FIT, BOUNDED BELOW AT Y=4

Projected data, 2000−2025

Year	WRD projection	ln (−4+col 2)	exp. proj. 2000
2000	5.7524998	0.5610432258	6.1107007471
2005	5.9718438	0.6789690441	5.703815248
2010	5.5329624	0.4272020725	5.3753661685
2015	5.0748488	0.0721800005	5.1102331075
2020	4.842618	−0.1712415672	4.8962104647
2025	4.69974	−0.3570464415	4.7234455463

Regression Output:

Constant	86.407026845
Std Err of Y Est	0.1351724445
R Squared	0.9165512002
No. of Observations	6
Degrees of Freedom	4
X Coefficient(s)	−0.04283487
Std Err of Coef.	0.0064624789

Linear equation:

$Ln(y−4) = −0.04283 * x + 86.40702$

Exponential equation to fit 2000 to 2025:

$y = @exp(−0.04283 * x + 86.40702) + 4$

7. The second equation, that is an exponential, is derived from the first equation by raising both sides to e, the base of the natural logarithms, and isolating y on the left side. This is the equation that will be used to enter projected exponential values, starting with the year 2000.

8. In a separate column of the spreadsheet, labelled "exp. proj. 2000" in Table 1.5 (Column D), enter the equation derived from the regression: in cell D5, enter the formula @EXP(-0.04283*A5+86.40702)+4. The value 6.1107007471 should appear at the top of the "exp. proj. 2000" column. (The number of decimal places displayed will depend upon the width of the column.) Then, copy the cell content from D5 to the 5 cells below it; this should produce the entire numerical range shown in the D column.

9. Graph the results; select an XY-graph. Put the 6 entries for the years in the X-range, and the six WRD projected values in the A range. Enter the 6 entries of the projected exponential data in an additional range, B, of the spreadsheet (Figure 1.7). The result should look like Figure 1.7.

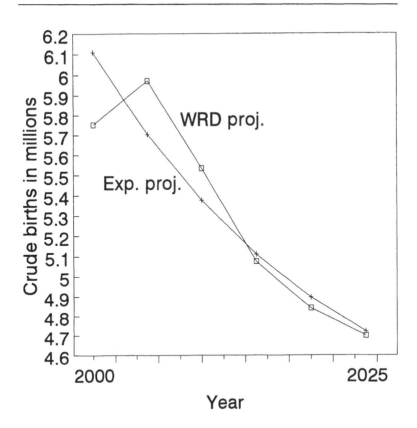

Figure 1.7. Graph of exponential fit to projected crude birth data, 2000-2025; the line y=4 is a horizontal asymptote for the exponential curve.

TABLE 1.6
Exponential curve fit to projected crude birth data, 2005-2025.
Horizontal asymptote at y=4.
(Source: World Resources Institute)

EXPONENTIAL FIT, BOUNDED BELOW AT Y=4

Projected data, 2005–2025

Year	WRD projection	exp. proj. 2000	exp. proj. 2005
2005	5.9718438	5.703815248	5.9800062475
2010	5.5329624	5.3753661685	5.516037465
2015	5.0748488	5.1102331075	5.1607890623
2020	4.842618	4.8962104647	4.8887849267
2025	4.69974	4.7234455463	4.6805186847

Regression Output:

Constant	107.750139444
Std Err of Y Est	0.0511744991
R Squared	0.9891033446
No. of Observations	5
Degrees of Freedom	3
X Coefficient(s)	−0.0534094922
Std Err of Coef.	0.0032365595

Linear equation:

$Ln(y-4) = -0.05340*x + 107.7501$

Exponential equation to fit 2005 to 2025:

$y = @exp(-0.05340*x + 107.7501) + 4$

The R-squared value for the 2000 exponential fit, with y=4 as a lower bound, is 0.916, which is closer to 1 than is the corresponding R-squared value for the 2000 exponential fit with y=0. Indeed, the graph in Figure 1.7 reflects not only this slight improvement, but more to the point, it shows the exponential fit tracking the WRD projections more closely; thus, one might ask whoever made the projections what sorts of assumptions they made concerning lower bounds and what basis in fact, plans, or policy led to the assumptions on which they made the lower bound choice.

Thus, one might expect an improved fit, reflected in the R-squared value, by fitting an exponential with a lower bound of y=4 to the data starting with the year 2005. Table 1.6 shows the results of performing that exponential fit. The R-squared is 0.989, an improvement over all previous fits. Indeed, when the WRD projection is graphed along with the two exponential projections using a lower bound of y=4, it is clear that the exponential beginning in 2005 tracks both the actual values of the WRD curve, as well as its general shape (Figure 1.8). One might try to fine-tune the fit even more; however, these fits are sufficient to direct some provocative questions with adequate visual support as back-up.

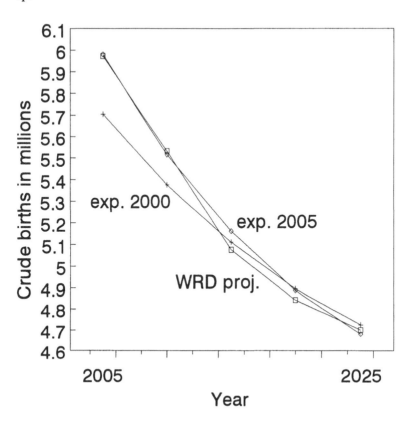

Figure 1.8. Graphs of exponential fits to projected crude birth data, 2000-2025, and 2005-2025; the line y=4 is a horizontal asymptote for each exponential curve.

BLACK BOX SUMMARY
see Introduction for theoretical explanation

EXPONENTIAL CURVE

$$y=Ce^{ax} + b$$
where
$$a < 0$$
and

y=b is the lower bound of the exponential;
C is a constant.

Logistic curve-fitting--Bangladesh total population data

Figure 1.1, which graphs total population data for Bangladesh using spreadsheet default curves, suggests that some sort of increasing curve might fit the actual total population data fairly well. Table 1.7 shows a straight line fit to that data, and Table 1.8 shows an exponential fit to that data. The straight line fit to the actual data is reasonable--r-squared is 0.978. The exponential fit is better--r-squared is 0.999. When the default curve, from the actual data, is graphed with the linear and the exponential fit for 1955-1990 (years of the actual data), it is also evident that the exponential offers a tight fit (Figure 1.9). One might suppose that projected values of total population were obtained using an exponential close to this one. When the WRD projections are graphed together with the linear and exponential projections, derived from extending the curves that actually fit the data, it is clear that some curve other than these was used for forecasting (Figure 1.10). The WRD projections lie between the linear and the exponential and the curve they trace out may well represent an exponential that has been damped by environmental factors, forcing it, eventually, to taper off horizontally. The curve that permits this sort of damping is the logistic curve--an S-shaped curve that exhibits exponential rise initially and then tapers off symmetrically around the center point of the "S," at which the slope of the curve (of the line tangent to the curve at that point) is steepest.

The mechanics of fitting a logistic curve to a table of values is straightforward to execute in a spreadsheet (again, reference is made to Lotus 1-2-3, release 2.3). A sequence of steps for doing so is presented below; all that is required is to select a starting point (usually a time, such as a year), an endpoint (again, often in time), an upper bound for the population (such as carrying capacity of the environment) and an assumption that the distribution be logistic in character.

TABLE 1.7
Least squares regression line fit to actual population data, 1955-1990.
Right-hand column shows fit of line to actual data
and extrapolation to cover 1995-2025.
(Source: World Resources Institute)

STRAIGHT LINE FIT			PROJECTED
Actual data, 1955–1990		Bangladesh	STRAIGHT
Year	Pop. mil.		LINE VALS.
1955	45.486		40.472895
1960	51.419		50.46124
1965	58.312		60.449585
1970	66.671		70.43793
1975	76.582		80.426275
1980	88.219		90.41462
1985	101.147		100.402965
1990	115.593		110.39131
1995	132.219		120.379655
2000	150.589		130.368
2005	170.138		140.356345
2010	188.196		150.34469
2015	204.631		160.333035
2020	220.119		170.32138
2025	234.987		180.309725

Regression Output:

Constant	−3864.9735714
Std Err of Y Est	3.9161593615
R Squared	0.9785117219
No. of Observations	8
Degrees of Freedom	6
X Coefficient(s)	1.9976690476
Std Err of Coef.	0.1208553017

Linear equation:
$y = 1.997669 * x - 3864.97$

TABLE 1.8
Exponential curve fit to actual and projected population data,
1955-2025.
Horizontal asymptote at y=0.
(Source: World Resources Institute)

EXPONENTIAL FIT			PROJECTED
Actual data, 1955–1990		Bangladesh	EXPONENTIAL
Year	Pop. mil.	Ln. Pop.	LINE VALS.
1955	45.486	3.8174045863	44.940077892
1960	51.419	3.940007754	51.406262317
1965	58.312	4.0658079041	58.802831
1970	66.671	4.1997700758	67.263651893
1975	76.582	4.3383620622	76.941854482
1980	88.219	4.4798223593	88.012601228
1985	101.147	4.6165749043	100.676257767
1990	115.593	4.7500754008	115.162019263
1995	132.219	4.8844596387	131.73205853
2000	150.589	5.0145542715	150.68627101
2005	170.138	5.1366098725	172.36770247
2010	188.196	5.2374839729	197.16875768
2015	204.631	5.3212083572	225.53830241
2020	220.119	5.3941683092	257.98978729
2025	234.987	5.4595301935	295.11054059

Regression Output:

Constant	−48.756876685
Std Err of Y Est	0.0077185033
R Squared	0.9995292746
No. of Observations	8
Degrees of Freedom	6
X Coefficient(s)	0.0268861114
Std Err of Coef.	0.0002381982

Linear equation:
LN y = 0.026886*x − 48.7568

Exponential equation:
y=@exp(0.026886*x − 48.7568)

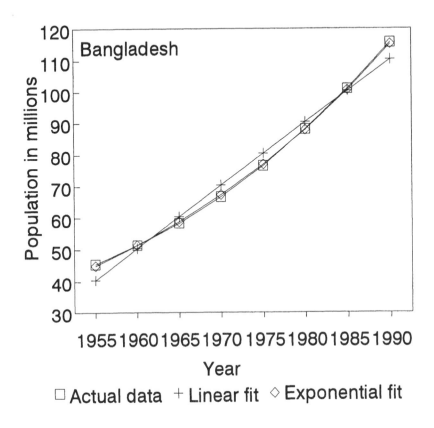

Figure 1.9. Linear and exponential curve fits to actual Bangladesh population data.

The general form for the logistic equation used here is:

$$y=q/(1+ae^{bx})$$

where b<0, q is selected prior to making any analysis and is the value of the upper bound selected by the user on carrying capacity or other bases, and a and b are constants to be determined depending on the values selected for q and the beginning and ending times chosen. The point of steepest slope is halfway between the lines y=0 and y=q; thus the choice of q influences directly where the steepest rate of increase in the curve occurs. There are numerous equivalent forms that this logistic equation might assume.

In the example below, there is data for the period from 1955 to 2025; the data from 1955 to 1990 is actual data, and that from 1995 to 2025 is projected. To see if a logistic was used to make the projection, we shall try to fit curves to all the data; since it is necessary only to choose one point as an end, it might well be that planners notice an exponential trend in actual data, as we have here, and, guessing that a purely exponential increase is not likely to remain, choose some value at a future time, as in 2025, that seems likely to them based on their accumulated wisdom and field evidence. The exponential part of the logistic will fit the earlier, exponentially rising years, and the years for projection, from the halfway point of the logistic, to the final year, are filled in by values projected from the curve. Whether or not this manner of forecasting is acceptable is a concern each time it is done; we illustrate with the case of total population in Bangladesh.

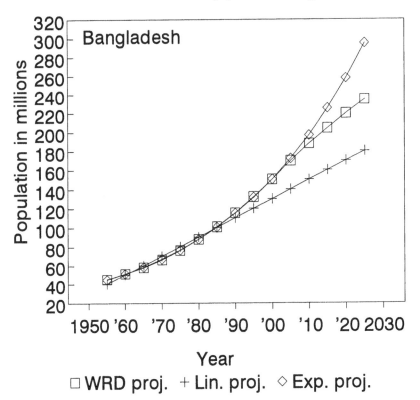

Figure 1.10. Curve showing WRD projections for population beyond 1990 to 2025. The linear fit and the exponential fit shown in Figure 1.9 are extrapolated here--the exponential overfits the WRD projections and the linear underfits the WRD projections.

LOGISTIC CURVES FIT TO THE DATA OF TABLE 1.9
(Refer to Table 1.9 and Figure 1.11)
Refer to Introduction for theoretical background

ASSUMPTIONS: Q=300; TIME PERIOD 1955-2025

1. Enter labels ranging from 0 to 14, one for each year, in column A--the first entry is in cell A5.
2. Enter the years for which there is data, both actual and projected--1955 to 2025 in this case--in column B.
3. Enter the data that varies over time in another column (spreadsheet column C--WRD projection in this case), to be used as y values.
4. The variable t will play the role of x in the logistic equation. To begin, assume that q=300 (million). Thus, the logistic equation is now $y=300(1+ae^{bt})$.
5. Find the constant a:
  In 1955, when t=0, y=45.486 (cell C5).
  Thus, the logistic equation now becomes, at that time,
  $45.486=300/(1+ae^{b*0})$, or more simply, $45.486=300/(1+a)$.
The right hand equation can be solved for the single variable a, by isolating that variable on the left as:
  $a=-1+300/45.486=5.595435$.
The logistic equation is now: $y=300/(1+5.595435e^{bt})$.
6. Now, find b--use information from the other endpoint selected--t=14, in 2025. When t=14, y=234.987. So, at this time, the logistic equation yields, substituting in values for y and t:
$234.987=300/(1+5.595435e^{14b})$.
This is an equation in a single variable; isolate the variable b on the left. Thus, $e^{14b}=(-1+(300/234.987))/5.595435$. The arithmetic of the right hand side reduces easily using a computer or calculator, so that $e^{14b}=0.049445$. Taking the natural logarithm of both sides produces $14b=\ln(0.049445)$ so that b=-0.21477. The b value has been determined and it is negative, as the theory tells us it must be. The logistic equation fit to the actual data is:
$y=300/(1+5.595435e^{-0.21477*t})$.

TABLE 1.9
Logistic curve fits to total population with various upper bounds of
q=300, q=350, q=375.
(Source: World Resources Institute)

LOGISTIC CURVE FITS TO DATA FROM 1955–2025

BANGLADESH, TOTAL POPULATION

$$y=q/(1+ae\,\hat{}\,(bx)), b<0$$

	Year	Pop. mil.	q=300	q=350	q=375
0	1955	45.486	45.486006609	45.486001683	45.48600523
1	1960	51.419	54.407079524	53.404581971	53.112302854
2	1965	58.312	64.633496693	62.419123561	61.777523509
3	1970	66.671	76.185803582	72.582920633	71.542254794
4	1975	76.582	89.021952777	83.918763407	82.444173779
5	1980	88.219	103.025274175	96.409750251	94.490087761
6	1985	101.147	117.999353994	109.991410623	107.648412342
7	1990	115.593	133.672861	124.546688921	121.843196948
8	1995	132.219	149.71566884	139.90531529	136.95093156
9	2000	150.589	165.76498412	155.84866999	152.80125632
10	2005	170.138	181.45742778	172.12042193	169.18228786
11	2010	188.196	196.46122567	188.44213133	185.85059338
12	2015	204.631	210.50263035	204.53193662	202.54501158
13	2020	220.119	223.3824358	220.1237297	219.00274704
14	2025	234.987	234.98119293	234.98410719	234.97567806

7. In a separate column of the spreadsheet, labelled "q=300" in Table 1.9 (Column D), enter the equation derived from the logistic equation: in cell D5, enter the formula 300/(1+5.95435*@EXP(-0.21477*A5)). The value 45.486006609 should appear at the top of the "q=300" column. (The number of decimal places displayed will depend upon the width of the column.) Then, copy the cell content from D5 to the 14 cells below it; this should produce the entire numerical range shown in the D column. The second value in the q=300 column is 64.633496693.

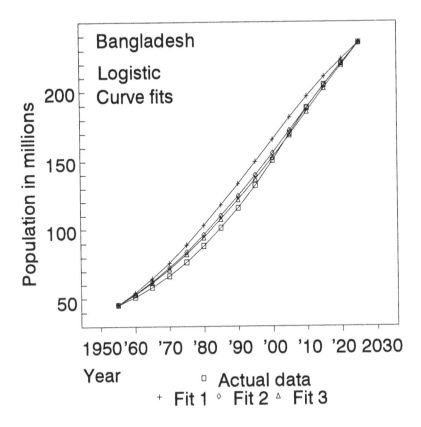

Figure 1.11. Logistic curve fits to total population in Bangladesh, projected and actual, from 1955-2025. In Fit 1, q=300; in Fit 2, q=350; in Fit 3, q=375.

8. Graph the results; select an XY-graph. Put the 14 entries for the years (column B) in the X-range, and the fourteen WRD projected values in the A range. Enter the 14 entries of the q=300 logistic data in an additional range, B, of the spreadsheet (Figure 1.11). The result should look like the curve labelled "Fit 1" in Figure 1.11.

Adjustment of the q value may afford an even closer fit; we illustrate the effects of doing so, first by taking q=350, and then by taking q=375. Table 1.9 shows the results of doing so, and these results are also graphed on Figure 1.11. The detail of the mechanics required to obtain these values is reproduced below.

LOGISTIC CURVES FIT TO THE DATA OF TABLE 1.9
(Refer to Table 1.9 and Figure 1.11)
Refer to Introduction for theoretical background

ASSUMPTIONS: Q=350; TIME PERIOD 1955-2025

1. Enter labels ranging from 0 to 14, one for each year, in
column A--the first entry is in cell A5.
2. Enter the years for which there is data, both actual and
projected--1955 to 2025 in this case--in column B.
3. Enter the data that varies over time in another column
(spreadsheet column C--WRD projection in this case), to be used
as y values.
4. The variable t will play the role of x in the logistic equation.
To begin, assume that q=350 (million).   Thus, the logistic
equation is now $y=350(1+ae^{bt})$.
5. Find the constant a:
   In 1955, when t=0, y=45.486 (cell C5).
   Thus, the logistic equation now becomes, at that time,
   $45.486=350/(1+ae^{b*0})$, or more simply, $45.486=350/(1+a)$.
The right hand equation can be solved for the single variable a,
by isolating that variable on the left as:
   a=-1+350/45.486=6.694675.
The logistic equation is now: $y=350/(1+6.694675e^{bt})$.
6.   Now, find b--use information from the other endpoint
selected--t=14, in 2025.   When t=14, y=234.987.   So, at this
time, the logistic equation yields, substituting in values for y and
t:
$234.987=350/(1+6.694675e^{14b})$.
This is an equation in a single variable; isolate the variable b on
the left.   Thus, $e^{14b}=(-1+(350/234.987))/6.694675$.   The
arithmetic of the right hand side reduces easily using a computer
or calculator, so that $e^{14b} = 0.0731095$ Taking the natural
logarithm of both sides produces 14b=ln(0.0731095) so that
b=-0.18684. The b value has been determined and it is negative,
as the theory tells us it must be. The logistic equation fit to the
actual data is:
$y=350/(1+6.694675e^{-0.18684*t})$.

7. In a separate column of the spreadsheet, labelled "q=350" in Table 1.9 (Column E), enter the equation derived from the logistic equation: in cell E5, enter the formula 350/(1+6.694675*@EXP(-0.18684*A5)). The value 45.486001683 should appear at the top of the "q=350" column. (The number of decimal places displayed will depend upon the width of the column.) Then, copy the cell content from E5 to the 14 cells below it; this should produce the entire numerical range shown in the E column. The second value in the q=350 column is 53.404581971.

8. Graph the results; select an XY-graph. Put the 14 entries for the years (column B) in the X-range, and the fourteen WRD projected values in the A range. Enter the 14 entries of the q=350 logistic data in an additional range, C, of the spreadsheet (Figure 1.11). The result should look like the curve labelled "Fit 2" in Figure 1.11.

The second logistic fit appears to fit the WRD data more tightly than did the first fit; thus, one might be tempted to increase the q value a bit more. When q=375, a third logistic curve is fit to the data; it appears to be a better fit than the first curve, but not as good a fit as the second curve (Table 1.9, Figure 1.11). The detail of the adjustment is given below.

LOGISTIC CURVES FIT TO THE DATA OF TABLE 1.9
(Refer to Table 1.9 and Figure 1.11)
Refer to Introduction for theoretical background

ASSUMPTIONS: Q=375; TIME PERIOD 1955-2025

1. Enter labels ranging from 0 to 14, one for each year, in column A--the first entry is in cell A5.

2. Enter the years for which there is data, both actual and projected--1955 to 2025 in this case--in column B.

3. Enter the data that varies over time in another column (spreadsheet column C--WRD projection in this case), to be used as y values.

4. The variable t will play the role of x in the logistic equation. To begin, assume that q=375 (million). Thus, the logistic equation is now $y=375(1+ae^{bt})$.

5. Find the constant a:
In 1955, when t=0, y=45.486 (cell C5).
Thus, the logistic equation now becomes, at that time,

$45.486 = 375/(1 + ae^{b*0})$, or more simply, $45.486 = 350/(1+a)$.
The right hand equation can be solved for the single variable a, by isolating that variable on the left as:
a=-1+375/45.486=7.244294.

The logistic equation is now: $y = 375/(1 + 7.244294e^{bt})$.
6. Now, find b--use information from the other endpoint selected--t=14, in 2025. When t=14, y=234.987. So, at this time, the logistic equation yields, substituting in values for y and t:

$234.987 = 375/(1 + 7.244294e^{14b})$.
This is an equation in a single variable; isolate the variable b on the left. Thus, $e^{14b} = (-1 + (375/234.987))/7.244294$. The arithmetic of the right hand side reduces easily using a computer or calculator, so that $e^{14b} = 0.0822486$. Taking the natural logarithm of both sides produces $14b = \ln(0.0822486)$ so that b=-0.17842. The b value has been determined and it is negative, as the theory tells us it must be. The logistic equation fit to the actual data is:

$y = 375/(1 + 7.244294e^{-0.17842*t})$.
7. In a separate column of the spreadsheet, labelled "q=375" in Table 1.9 (Column F), enter the equation derived from the logistic equation: in cell F5, enter the formula 375/(1+7.244294*@EXP(-0.17842*A5)). The value 45.48600523 should appear at the top of the "q=375" column. (The number of decimal places displayed will depend upon the width of the column.) Then, copy the cell content from F5 to the 14 cells below it; this should produce the entire numerical range shown in the F column. The second value in the q=375 column is 53.112302854.
8. Graph the results; select an XY-graph. Put the 14 entries for the years (column B) in the X-range, and the fourteen WRD projected values in the A range. Enter the 14 entries of the q=375 logistic data in an additional range, D, of the spreadsheet (Figure 1.11). The result should look like the curve labelled "Fit 3" in Figure 1.11.

Logistic curves have often been used as models to describe the diffusion of biological, physical, and even social phenomena. In the latter vein, Törsten Hägerstrand (a Swedish geographer at the University of Lund) employed Monte Carlo simulation, and logistic curves, to study the diffusion of an innovation within a population; the results of his elegant study suggest that, generally speaking, people tend not to move far from their origins (see references). This result also fits with common sense, as we reflect on the perpetuation of cultural enclaves as well as on the perhaps surprisingly high percentage of our classmates who remain near their secondary schools.

BLACK BOX SUMMARY
see Introduction for theoretical explanation

LOGISTIC CURVE
$$y=q/(1+ae^{bx})$$
where
$$b < 0$$
and
y=q is the upper bound of the curve, chosen ahead of time;
a and b are constants, calculated from the data.

Gompertz curve--Bangladesh total population
    The process of fitting a logistic curve to a set of data is quite different from the process of fitting a straight line, using least squares, or an exponential to the same data set. With the logistic fit, two points and a line (the q value) control the shape of the S; indeed, the q-value functions as a ceiling pressing down on exponential growth, much as a living room ceiling inhibits the "natural" growth of tall indoor plants. We might wonder, then, if there are various modifications of the logistic curve that allow for creating S-shapes using a variation in how the ceiling value is chosen.

    One variant of the logistic curve, in which the S-shape appears flatter is the Gompertz curve; it is used to model growth of various kinds, from financial to population. The reason the curve is flatter become evident when the logistic equation is written as a differential equation, dP/dt = P(a-b*P), and the Gompertz is also written in an equivalent manner, as dP/dt = P(a-b*ln P)--the logarithmic function tends to flatten out the curve and make the S-shape less curved than would a logistic fit.

Gompertz curves are fit to the Bangladesh total population values, assuming upper bounds of 300, 350, and 375 million (Table 1.10); the graphs of these fits are shown in Figure 1.12. A brief glance at these curves shows that they all lie above the actual data and that they are too straight--the Gompertz is not as good a choice as a logistic curve in fitting this data. One of the assumptions in fitting S-shaped curves is on the upper bound; another, in the previous case, is that the curve follow a logistic equation. Here, we alter the logistic assumption and fit curves to the same years, population values, and upper bounds. The mechanics of fitting the curves is given in the sequence of steps below.

The general form for the Gompertz equation used here is:

$$y = q * e^{-ce^{-bx}}$$

where q is selected prior to making any analysis and is the value of the upper bound selected by the user on carrying capacity or other bases, and b and c are constants to be determined depending of the values selected for q and the beginning and ending times chosen. There are numerous equivalent forms that this Gompertz equation might assume.

GOMPERTZ CURVES FIT TO THE DATA OF TABLE 1.10
(Refer to Table 1.10 and Figure 1.12)
Refer to Introduction for theoretical background

ASSUMPTIONS: Q=300; TIME PERIOD 1955-2025

1. Enter labels ranging from 0 to 14, one for each year, in column A--the first entry is in cell A5.
2. Enter the years for which there is data, both actual and projected--1955 to 2025 in this case--in column B.
3. Enter the data that varies over time in another column (spreadsheet column C--WRD projection in this case), to be used as y values.
4. The variable t will play the role of x in the Gompertz equation. To begin, assume that q=300 (million). Thus, the Gompertz equation is now

$y = 300 * e^{-ce^{-bx}}$.

TABLE 1.10
Gompertz curve fits to total population with various upper bounds of q=300, q=350, q=375.
(Source: World Resources Institute)

GOMPERTZ CURVE FITS TO DATA FROM 1955–2025

BANGLADESH, TOTAL POPULATION

$$y=qe\,\hat{}\,(-ce\,\hat{}\,(bx))$$

	Year	Pop. mil.	q=300	q=350	q=375
0	1955	45.486	45.48599947	45.485998553	45.486001804
1	1960	51.419	58.772805652	56.947262424	56.412479021
2	1965	58.312	73.342308251	69.553650985	68.443276262
3	1970	66.671	88.810926489	83.100309853	81.417612618
4	1975	76.582	104.782084579	97.358595139	95.150770509
5	1980	88.219	120.878884778	112.091378638	109.445471328
6	1985	101.147	136.767278	127.06632436	124.102308739
7	1990	115.593	152.16956413	142.06637493	138.92857804
8	1995	132.219	166.86954786	156.89720057	153.74516315
9	2000	150.589	180.71136562	171.39174707	168.39141074
10	2005	170.138	193.59409674	185.41226275	182.7281167
11	2010	188.196	205.46401612	198.85030525	196.63887384
12	2015	204.631	216.30593386	211.6252556	210.03009058
13	2020	220.119	226.13463315	223.68183117	222.83000419
14	2025	234.987	234.98703537	234.98701751	234.98699227

5. Find the constant c:
   In 1955, when t=0, y=45.486 (cell C5).
   Thus, the logistic equation now becomes, at that time,
   $$45.486=300*e^{-c}.$$
The right hand equation can be solved for the single variable c, by isolating that variable on the left as:
   $$c=\ln(300/45.486)=1.8863779.$$
The Gompertz equation is now:
   $$y=300*e^{-1.8863779e^{-bx}}.$$

6.   Now, find b--use information from the other endpoint selected--t=14, in 2025.   When t=14, y=234.987.   So, at this time, the Gompertz equation yields, substituting in values for y and t:

$$234.987=300*e^{-1.8863779e^{-14b}}.$$

This is an equation in a single variable; isolate the variable b on the left.   Thus, $-1.8863779e^{-14b}$=ln(234.987/300)=-0.2442523. Thus, $0.1294822=e^{-14b}$ so   that   -b   =   1/14 ln(0.2442523/1.8863779)   =   -0.1460152.   The   Gompertz equation fit to the actual data is:

$$y=300*e^{-1.8863779e^{-0.1460152x}}$$

7.   In a separate column of the spreadsheet, labelled "q=300" in Table 1.10 (Column D), enter the equation derived from the Gompertz equation: in   cell   D5,   enter   the   formula 300*(@EXP(-1.8863779*@EXP(-0.1460152*A5))).   The value 45.48599947 should appear at the top of the "q=300" column. (The number of decimal places displayed will depend upon the width of the column.)   Then, copy the cell content from D5 to the 14 cells below it; this should produce the entire numerical range shown in the D column.   The second value in the q=300 column is 58.77805652 (less than the corresponding logistic value, and closer to the WRD value).

8. Graph the results; select an XY-graph.   Put the 14 entries for the years (column B) in the X-range, and the fourteen WRD projected values in the A range.   Enter the 14 entries of the q=300 Gompertz data in an additional range, B, of the spreadsheet (Figure 1.12). The result should look like the curve labelled "Fit 1" in Figure 1.12.

Adjustment of the q value may afford an even closer fit; we illustrate the effects of doing so, first by taking q=350, and then by taking q=375. Table 1.10 shows the results of doing so, and these results are also graphed on Figure 1.12. The detail of the mechanics required to obtain these values is reproduced below.

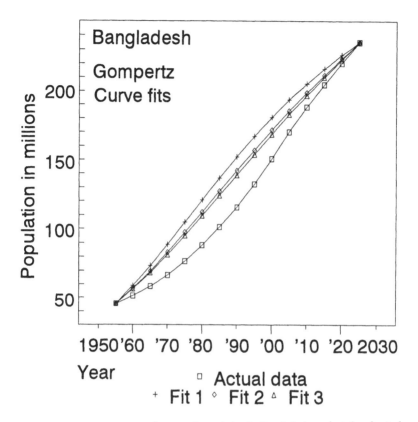

Figure 1.12. Gompertz curve fits to total population in Bangladesh, projected and actual, from 1955-2025.   In Fit 1, q=300; in Fit 2, q=350; in Fit 3, q=375.

GOMPERTZ CURVES FIT TO THE DATA OF TABLE 1.10
(Refer to Table 1.10 and Figure 1.12)
Refer to Introduction for theoretical background

ASSUMPTIONS: Q=350; TIME PERIOD 1955-2025

1.   Enter labels ranging from 0 to 14, one for each year, in column A--the first entry is in cell A5.
2.   Enter the years for which there is data, both actual and projected--1955 to 2025 in this case--in column B.
3.   Enter the data that varies over time in another column (spreadsheet column C--WRD projection in this case), to be used as y values.
4.   The variable t will play the role of x in the Gompertz equation. To begin, assume that q=350 (million). Thus, the Gompertz

equation is now

$$y=350*e^{-ce^{-bx}} .$$

5. Find the constant c:

In 1955, when t=0, y=45.486 (cell C5).

Thus, the logistic equation now becomes, at that time,

$$45.486=350*e^{-c}.$$

The right hand equation can be solved for the single variable c, by isolating that variable on the left as:

c=ln(350/45.486)=2.0405286.

The Gompertz equation is now:

$$y=350*e^{-2.0405286e^{-bx}} .$$

6. Now, find b--use information from the other endpoint selected--t=14, in 2025. When t=14, y=234.987. So, at this time, the Gompertz equation yields, substituting in values for y and t:

$$234.987=300*e^{-2.0405286e^{-14b}} .$$

This is an equation in a single variable; isolate the variable b on the left. Thus, $-2.0405286e^{-14b}$=ln(234.987/350)=-0.398403. Thus, -b = 1/14 ln(0.398403/2.0405286) = -0.1166786. The Gompertz equation fit to the actual data is:

$$y=350*e^{-2.0405286e^{-0.1166786x}}$$

7. In a separate column of the spreadsheet, labelled "q=350" in Table 1.10 (Column E), enter the equation derived from the Gompertz equation: in cell E5, enter the formula 350*(@EXP(-2.0405286*@EXP(-0.1166786*A5))). The value 45.485998553 should appear at the top of the "q=350" column. (The number of decimal places displayed will depend upon the width of the column.) Then, copy the cell content from E5 to the 14 cells below it; this should produce the entire numerical range shown in the E column. The second value in the q=350 column is 56.947262424 (less than the corresponding logistic value, and closer to the WRD value).

8. Graph the results; select an XY-graph. Put the 14 entries for the years (column B) in the X-range, and the fourteen WRD projected values in the A range. Enter the 14 entries of the q=350 Gompertz data in an additional range, C, of the spreadsheet (Figure 1.12). The result should look like the curve labelled "Fit 2" in Figure 1.12.

GOMPERTZ CURVES FIT TO THE DATA OF TABLE 1.10
(Refer to Table 1.10 and Figure 1.12)
Refer to Introduction for theoretical background

ASSUMPTIONS: Q=375; TIME PERIOD 1955-2025

1. Enter labels ranging from 0 to 14, one for each year, in column A--the first entry is in cell A5.
2. Enter the years for which there is data, both actual and projected--1955 to 2025 in this case--in column B.
3. Enter the data that varies over time in another column (spreadsheet column C--WRD projection in this case), to be used as y values.
4. The variable t will play the role of x in the Gompertz equation. To begin, assume that q=375 (million). Thus, the Gompertz equation is now

$y=375*e^{-ce^{-bx}}$.

5. Find the constant c:
In 1955, when t=0, y=45.486 (cell C5).
Thus, the logistic equation now becomes, at that time,

$45.486=375*e^{-c}$.

The right hand equation can be solved for the single variable c, by isolating that variable on the left as:

$c=\ln(375/45.486)=2.1095214$.

The Gompertz equation is now:

$$y=375*e^{-2.2.1095214e^{-bx}}$$.

6. Now, find b--use information from the other endpoint selected--t=14, in 2025. When t=14, y=234.987. So, at this time, the Gompertz equation yields, substituting in values for y and t:

$$234.987=300*e^{-2.21095214e^{-14b}}$$.

This is an equation in a single variable; isolate the variable b on the left. Thus, $-2.21095214e^{-14b}=\ln(234.987/375)=-0.4673958$. Thus, $-b = 1/14 \ln(0.4673958/2.1095214) = -0.1076457$. The Gompertz equation fit to the actual data is:

$$y=375*e^{-2.2109521e^{-0.1076457t}}$$.

7. In a separate column of the spreadsheet, labelled "q=375" in Table 1.10 (Column F), enter the equation derived from the Gompertz equation: in cell F5, enter the formula 375*(@EXP(-2.2109521*@EXP(-0.1076457*A5))). The value 45.486001804 should appear at the top of the "q=375" column. (The number of decimal places displayed will depend upon the width of the column.) Then, copy the cell content from F5 to the 14 cells below it; this should produce the entire numerical range shown in the F column. The second value in the q=375 column is 56.412479021 (just slightly less than the corresponding logistic value, and closer to the WRD value).

8. Graph the results; select an XY-graph. Put the 14 entries for the years (column B) in the X-range, and the fourteen WRD projected values in the A range. Enter the 14 entries of the q=375 Gompertz data in an additional range, D, of the spreadsheet (Figure 1.12). The result should look like the curve labelled "Fit 3" in Figure 1.12.

These Gompertz curves appear generally not to fit the WRD data as well as do the logistic curves; however, the Gompertz fit is better to the early data, where the increase is steep--then, the relative flatness of the Gompertz curves carries them, from a steep start, to overfit the later WRD values. Again, these observations suggest the importance of understanding the various ways projections might be made--apparently subtle differences can translate into major differences when regional policies, involving large projects and sums of investment, are based on forecasts.

BLACK BOX SUMMARY

GOMPERTZ CURVE

$$y = q * e^{-ce^{-bt}}$$

where
y=q is the upper bound of the curve, chosen ahead of time;
b and c are constants, calculated from the data.

Total population and crude births

In looking at the relationship between total population and total crude births, a starting point might be to fit a straight line to "total population" (independent variable) and "total crude births" (dependent variable). Table 1.11 shows the needed data; when the regression feature of Lotus 1-2-3 is used to fit a straight line to the actual data, from 1955 to 1990, a good fit is established, with R-squared 0.9797306479. The right hand column uses the linear equation derived from the regression, noted at the bottom of Table 1.11, to project values for the crude birth variable from the population variable. The linear data closely track the actual WRD data as is reflected in the left-hand side of the Figure 1.13. When the linear fit is projected to the future, it is different from the WRD projection; indeed, the R-squared value for the fit of all the data, 1955-2025, is 0.594527, and the R-squared value for the fit of the projected data, alone (from 1990-2025), is 0.141413!

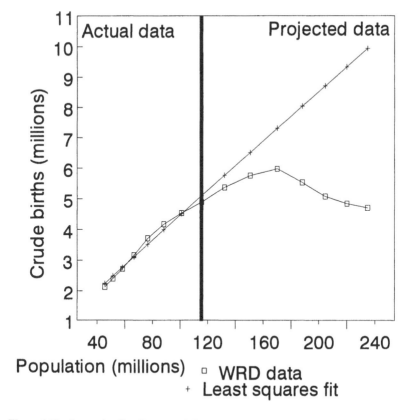

Figure 1.13. Regression line fit to actual data , 1955-1990, for Bangladesh. Linear and WRD projections appear to the right of the heavy vertical line.

TABLE 1.11
(Source: World Resources Institute)

STRAIGHT LINE FIT			PROJECTED
Actual data, 1955–1990		Bangladesh	STRAIGHT
Year	Pop. mil.	Crude births, mil	LINE VALS.
1955	45.486	2.137842	2.2494626234
1960	51.419	2.406409	2.4903542894
1965	58.312	2.723170	2.7702238754
1970	66.671	3.166872	3.1096159934
1975	76.582	3.714227	3.5120224154
1980	88.219	4.163936	3.9845078894
1985	101.147	4.531385	4.5094105454
1990	115.593	4.878024	5.0959470374
1995	132.219	5.368091	5.7709958894
2000	150.589	5.752499	6.5168546294
2005	170.138	5.971843	7.3105831274
2010	188.196	5.532962	8.0437740434
2015	204.631	5.074848	8.7110679134
2020	220.119	4.842618	9.3399116894
2025	234.987	4.69974	9.9435822254

Regression Output:

Constant	0.4026400514
Std Err of Y Est	0.1560196484
R Squared	0.9797306479
No. of Observations	8
Degrees of Freedom	6
X Coefficient(s)	0.040602536
Std Err of Coef.	0.0023842079

Linear equation:
$y = 0.040602 * x + 0.4026400514$

A good fit of actual data by one function does not necessarily mean that that function will be used to make projections; there are an infinite number of functions that fit any finite display of data points. In this case, the background work suggests that the unexpected drop occurs in the projected birth data and that therefore one might need to do further research to determine why the decline is projected. Often it is helpful to separate data sets in order to understand why particular patterns occur.

References

1. Bogue, D. J. *Principles of Demography*, Wiley, New York, 1969.
2. Freedman, D., Pisani, R., Purves, R. *Statistics*, W. W. Norton, New York, 1978.
3. Hägerstrand, T. *Innovation Diffusion as a Spatial Process*, University of Chicago Press, Chicago, 1967.
4. Keyfitz, N. *Introduction to the Mathematics of Population*, Addison-Wesley, Reading, Massachusetts, 1968.
5. Ness, G. D., Drake, W. D., Brechin, S. R. (eds.) *Population-Environment Dynamics: Ideas and Observations*, The University of Michigan Press, Ann Arbor, Michigan, 1993.
6. Zill, D. G. *A First Course in Differential Equations* (fifth edition), Boston, PWS-Kent, 1993.

CHAPTER 2

EPIDEMIOLOGY DATA ANALYSIS

ANALYTICAL TECHNIQUES/TOOLS USED:

Consistent database construction
Mapping of data
Straight line curve-fitting--least squares
Root mean square error

DATA TYPE: ABUNDANT BASELINE DATA

cleaning, analysis, graphing, and mapping of data

Overview of Data

Epidemiology is a science that considers the incidence, distribution, and control of disease in human populations. One way to control disease is through immunization programs for children. The World Resources Institute data base has various indicators that are related to epidemiological issues; we consider the four for which there is data on the immunization of 1-year-old children against tuberculosis (TB), diphtheria-pertussis-tetanus (DPT), polio, and measles. Data is available for many countries of the world, but only for the year 1990. Techniques of curve fitting, such as those often used in population studies for which there is abundant longitudinal data, are not applicable in situations of this sort. Creative use of various tools is required in order to gain information from data sets that are limited in one way or another.

With any data set (presented in electronic or paper format), it is important first to examine the set for interesting or unusual patterns in the display. These patterns often influence decisions in choosing subsets of data and tools to analyze subsets.

61

PATTERNS IN DATA--WHAT TO LOOK FOR

1. What is the general organizational scheme of the entire set? Is it arranged alphabetically, numerically, or in some other fashion?

2. Are the real-world entries in the Table (nations, states, counties) expressed as comparable units? For example, county data and national data are generally not comparable.

3. Are the numerical entries in the Table expressed in comparable units? For example, data in one column might measure percentages while data in another column might measure thousands of dollars--these columns would not be comparable.

4. Are there gaps in the data? If so, what is their significance to the questions you wish to have the data answer?

Table 2.1 shows the results of transferring data directly from the WRD data base to a spreadsheet. The data has not been adjusted in any significant way (merely resized and rearranged to fit the page requirements of this book). Consider this table with respect to each of the four points above.

1. What is the general organizational scheme of the entire set? Is it arranged alphabetically, numerically, or in some other fashion?

Table 2.1 is arranged in alphabetical order; that is how it was exported from the data base to the spreadsheet. There are also columns of numbers, so that once the data is in the spreadsheet, it could be rearranged numerically.

2. Are the real-world entries in the Table (nations, states, counties) expressed as comparable units? For example, county data and national data are generally not comparable.

There are units at three different geographical scales in Table 2.1. Most of the data is at the national level; there are two entries for each variable at the continental scale, and one entry for each variable at the global scale. The continental and global entries are noted in upper case letters.

TABLE 2.1 (source: World Resources Institute)

Immunization of 1−year−olds   Percent, 1990.

Country	TB	DPT	Polio	Measles
Afghanistan	30 Afghanistan	25 Afghanistan	25 Afghanistan	20
Algeria	99 Algeria	89 Algeria	89 Algeria	83
Angola	47 Angola	23 Angola	23 Angola	38
Argentina	99 Argentina	85 Argentina	89 Argentina	95
Bangladesh	86 Bahrain	95 Bahrain	95 Bahrain	86
Barbados	95 Bangladesh	62 Bangladesh	62 Bangladesh	54
Bhutan	99 Barbados	91 Barbados	90 Barbados	87
Bolivia	48 Bhutan	95 Bhutan	95 Bhutan	89
Botswana	92 Bolivia	41 Bolivia	50 Bolivia	53
Brazil	78 Botswana	86 Botswana	82 Botswana	78
Belize	80 Brazil	81 Brazil	93 Brazil	78
Solomon Is.	87 Belize	84 Belize	80 Belize	81
Myanmar	75 Solomon Is.	77 Solomon Is.	75 Solomon Is.	70
Burundi	97 Myanmar	69 Myanmar	69 Myanmar	73
Cambodia	54 Burundi	86 Burundi	86 Burundi	75
Cameroon	76 Cambodia	40 Cambodia	40 Cambodia	34
Cape Verde	97 Cameroon	56 Cameroon	54 Cameroon	56
C. Afr. Rep.	96 Cape Verde	88 Cape Verde	87 Cape Verde	79
Sri Lanka	88 C. Afr. Rep.	82 C. Afr. Rep.	82 C. Afr. Rep.	82
Chad	59 Sri Lanka	90 Sri Lanka	90 Sri Lanka	83
Chile	97 Chad	20 Chad	20 Chad	32
China	99 Chile	99 Chile	99 Chile	98
Colombia	95 China	97 China	98 China	98
Comoros	99 Colombia	87 Colombia	93 Colombia	82
Congo	90 Comoros	94 Comoros	94 Comoros	87
Zaire	65 Congo	79 Congo	79 Congo	75
Costa Rica	92 Zaire	32 Zaire	31 Zaire	31
Cuba	98 Costa Rica	95 Costa Rica	95 Costa Rica	90
Benin	92 Cuba	92 Cuba	94 Cuba	94
Dom. Rep.	68 Cyprus	90 Cyprus	90 Cyprus	74
Ecuador	88 Benin	67 Benin	67 Benin	70
El Salvador	60 Dom. Rep.	69 Dom. Rep.	90 Dom. Rep.	96
Eq. Guinea	97 Ecuador	68 Ecuador	67 Ecuador	61
Ethiopia	56.96 El Salvador	76 El Salvador	76 El Salvador	75
Fiji	99 Eq. Guinea	78 Eq. Guinea	75 Eq. Guinea	88
Djibouti	95 Ethiopia	44.16 Ethiopia	44.16 Ethiopia	37.12
Gabon	96 Fiji	97 Fiji	96 Fiji	84
Gambia	99 Djibouti	85 Djibouti	85 Djibouti	85
Ghana	81 Gabon	78 Gabon	78 Gabon	76
Guatemala	62 Gambia	90 Gambia	93 Gambia	73
Guinea	53 Ghana	57 Ghana	56 Ghana	60
Guyana	85 Guatemala	66 Guatemala	74 Guatemala	68

Haiti	72 Guinea	17 Guinea	17 Guinea	18
Honduras	71 Guyana	83 Guyana	79 Guyana	73
India	97 Haiti	41 Haiti	40 Haiti	31
Indonesia	93 Honduras	84 Honduras	87 Honduras	90
Iran, Islamic	95 India	92 India	93 India	87
Iraq	96 Indonesia	87 Indonesia	91 Indonesia	86
Cote d'Ivoire	63 Iran, Islamic	93 Iran, Islamic	92 Iran, Islamic	83
Jamaica	98 Iraq	75 Iraq	75 Iraq	62
Kenya	80 Cote d'Ivoire	48 Cote d'Ivoire	48 Cote d'Ivoire	42
Korea, DPR	99 Jamaica	86 Jamaica	87 Jamaica	74
Korea, Rep	72 Jordan	92 Jordan	92 Jordan	87
Lao PDR	26 Kenya	74 Kenya	71 Kenya	59
Lesotho	97 Korea, DPR	98 Korea, DPR	99 Korea, DPR	99
Liberia	62 Korea, Rep	74 Korea, Rep	74 Korea, Rep	95
Libya	90 Kuwait	94 Kuwait	94 Kuwait	98
Madagascar	67 Lao PDR	18 Lao PDR	26 Lao PDR	13
Malawi	97 Lebanon	82 Lebanon	82 Lebanon	39
Malaysia	99 Lesotho	76 Lesotho	75 Lesotho	76
Mali	82 Liberia	28 Liberia	28 Liberia	55
Mauritania	75 Libya	84 Libya	84 Libya	70
Mauritius	94 Madagascar	46 Madagascar	46 Madagascar	33
Mexico	70 Malawi	81 Malawi	79 Malawi	80
Mongolia	92 Malaysia	91 Malaysia	90 Malaysia	90
Morocco	96 Mali	42 Mali	42 Mali	43
Mozambique	59 Mauritania	28 Mauritania	28 Mauritania	33
Oman	93 Mauritius	90 Mauritius	90 Mauritius	84
Namibia	85 Mexico	66 Mexico	96 Mexico	78
Nepal	97 Mongolia	84 Mongolia	85 Mongolia	86
Nicaragua	81 Morocco	81 Morocco	81 Morocco	79
Niger	50 Mozambique	46 Mozambique	46 Mozambique	58
Nigeria	96 Oman	96 Oman	96 Oman	96
Pakistan	98 Namibia	53 Namibia	53 Namibia	41
Panama	83 Nepal	79 Nepal	78 Nepal	67
Papua	89 Nicaragua	65 Nicaragua	86 Nicaragua	82
Paraguay	90 Niger	13 Niger	13 Niger	21
Peru	83 Nigeria	57 Nigeria	57 Nigeria	54
Philippines	96 Pakistan	96 Pakistan	96 Pakistan	97
Guinea–Bis	90 Panama	82 Panama	82 Panama	99
Qatar	97 Papua	69 Papua	69 Papua	67
Rwanda	92 Paraguay	78 Paraguay	76 Paraguay	69
Saudi Arabia	99 Peru	72 Peru	73 Peru	64
Senegal	92 Philippines	88 Philippines	88 Philippines	85
Sierra Leone	98 Guinea–Bis	38 Guinea–Bis	38 Guinea–Bis	42
Singapore	99 Qatar	82 Qatar	82 Qatar	79
Viet Nam	90 Rwanda	84 Rwanda	83 Rwanda	83
Somalia	31 Saudi Arabia	94 Saudi Arabia	94 Saudi Arabia	90
Zimbabwe	71 Senegal	60 Senegal	66 Senegal	59
Sudan	73 Sierra Leone	83 Sierra Leone	83 Sierra Leone	75
Swaziland	96 Singapore	85 Singapore	85 Singapore	87
Syrian Arab	92 Viet Nam	87 Viet Nam	87 Viet Nam	87
Thailand	99 Somalia	18 Somalia	18 Somalia	30
Togo	94 Zimbabwe	73 Zimbabwe	72 Zimbabwe	69

U. Arab E.	96	Sudan	62	Sudan	62	Sudan	57
Tunisia	99	Suriname	83	Suriname	81	Suriname	65
Uganda	99	Swaziland	89	Swaziland	89	Swaziland	85
Egypt	88	Syrian Arab	90	Syrian Arab	90	Syrian Arab	87
Tanzania	93	Thailand	92	Thailand	92	Thailand	80
Burkina F.	84	Togo	61	Togo	61	Togo	57
Uruguay	99	Trinidad	83	Trinidad	83	Trinidad	69
Venezuela	63	U. Arab E.	85	U. Arab E.	85	U. Arab E.	75
Yemen AR	99	Tunisia	90	Tunisia	90	Tunisia	87
Zambia	97	Turkey	84	Turkey	84	Turkey	78
WORLD	90	Uganda	77	Uganda	77	Uganda	74
AFRICA	80	Egypt	87	Egypt	87	Egypt	86
ASIA	95	Tanzania	85	Tanzania	82	Tanzania	83
		Burkina F.	37	Burkina F.	37	Burkina F.	42
		Uruguay	88	Uruguay	88	Uruguay	82
		Venezuela	63	Venezuela	72	Venezuela	62
		Yemen AR	89	Yemen AR	89	Yemen AR	74
		Zambia	79	Zambia	78	Zambia	76
		WORLD	83	WORLD	85	WORLD	80
		AFRICA	56	AFRICA	55	AFRICA	54
		ASIA	90	ASIA	91	ASIA	88

3.   Are the numerical entries in the Table expressed in comparable units? For example, data in one column might measure percentages while data in another column might measure thousands of dollars--these columns would not be comparable.

All the entries in Table 2.1 are expressed as percentages.

4. Are there gaps in the data? If so, what is their significance to the questions you wish to have the data answer?

There are gaps in this data set. If there were no gaps, each country name would be the same in any single line, throughout the entire expanse of Table 2.1.   In the first line, each country name is Afghanistan; in the second it is Algeria; in the third it is Angola; in the fourth it is Argentina; however, it the fifth, the first column is for Bangladesh, while subsequent entries in the same line are for Bahrain data. Clearly there is a gap here. If it is not considered, there is the risk that all data from this point on may become associated with incorrect countries. Were that to happen, all analyses based on that data would be wrong and, if allowed to stand, might result in very serious misallocation of funds.

Consistent database construction by elimination

Thus we consider, at length, strategies for constructing consistent data sets from material derived from publicly-available data bases. In Table 2.1, the first column is of shorter total length that are the other three which are of equal length (compare the positions of the last entries in each column). It would appear that the first column differs from the others in at least 8 positions; and, it may be the case that the other three columns are identical. It makes sense to check on the easiest possibilities first. Indeed, a glance at the country names for the variables DPT, Polio, and Measles, shows that each has the same set of countries associated with it. Only the set of countries associated with the TB variable differs. This simple observation means that if one wishes to generate a consistent data set from Table 2.1, then, as one possibility, all the countries in the larger set can be retained only if one considers three variables, DPT, Polio, and Measles. The data on TB would be omitted. Table 2.2 shows a consistent data set derived in this manner from Table 2.1.

Because all the entries are retained, it would make sense to compare individual national percentages to the global percentages for the entire set. In a truncated set, it would not make sense to compare individual national values to global values derived from a different set of data. It is difficult to use a data set that is as long as this one to make comparisons of national to global percentages. A bar chart of the sort shown in Figure 2.1 offers an effective visual display for doing so; however, this sort of display is necessarily limited to a small number of countries at a time.

In Figure 2.1, the spreadsheet was used to create a bar chart for each of eight countries, with three adjacent bars used for each country to represent each of the three diseases. Horizontal lines were added at the percentage level of the World data for each of the three diseases. The global rate for polio immunization of 1-year-olds is 85%, for DPT immunization it is 83%, and for measles it is 80%. It is then easy to see which countries fall above or below the global figures and whether or not they do so in all three immunization programs or only in one or two. A great deal of information can be transmitted using variables at different geographic scales (national and global in this case). Diagrams of this sort can be used in a variety of contexts.

This strategy of mixing scales can be applied independent of the existence of longitudinal data--it is a strategy that is spatial rather than temporal. In the absence of temporal data, one is forced, therefore, to rely solely on spatial analyses of data.

TABLE 2.2 (source: World Resources Institute)

Immunization of 1-year-olds   Percent, 1990.

	DPT		Polio		Measles
Afghanistan	25	Afghanistan	25	Afghanistan	20
Algeria	89	Algeria	89	Algeria	83
Angola	23	Angola	23	Angola	38
Argentina	85	Argentina	89	Argentina	95
Bahrain	95	Bahrain	95	Bahrain	86
Bangladesh	62	Bangladesh	62	Bangladesh	54
Barbados	91	Barbados	90	Barbados	87
Bhutan	95	Bhutan	95	Bhutan	89
Bolivia	41	Bolivia	50	Bolivia	53
Botswana	86	Botswana	82	Botswana	78
Brazil	81	Brazil	93	Brazil	78
Belize	84	Belize	80	Belize	81
Solomon Is.	77	Solomon Is.	75	Solomon Is.	70
Myanmar	69	Myanmar	69	Myanmar	73
Burundi	86	Burundi	86	Burundi	75
Cambodia	40	Cambodia	40	Cambodia	34
Cameroon	56	Cameroon	54	Cameroon	56
Cape Verde	88	Cape Verde	87	Cape Verde	79
C. Afr. Rep.	82	C. Afr. Rep.	82	C. Afr. Rep.	82
Sri Lanka	90	Sri Lanka	90	Sri Lanka	83
Chad	20	Chad	20	Chad	32
Chile	99	Chile	99	Chile	98
China	97	China	98	China	98
Colombia	87	Colombia	93	Colombia	82
Comoros	94	Comoros	94	Comoros	87
Congo	79	Congo	79	Congo	75
Zaire	32	Zaire	31	Zaire	31
Costa Rica	95	Costa Rica	95	Costa Rica	90
Cuba	92	Cuba	94	Cuba	94
Cyprus	90	Cyprus	90	Cyprus	74
Benin	67	Benin	67	Benin	70
Dom. Rep.	69	Dom. Rep.	90	Dom. Rep.	96
Ecuador	68	Ecuador	67	Ecuador	61
El Salvador	76	El Salvador	76	El Salvador	75
Eq. Guinea	78	Eq. Guinea	75	Eq. Guinea	88
Ethiopia	44.16	Ethiopia	44.16	Ethiopia	37.12
Fiji	97	Fiji	96	Fiji	84
Djibouti	85	Djibouti	85	Djibouti	85
Gabon	78	Gabon	78	Gabon	76
Gambia	90	Gambia	93	Gambia	73
Ghana	57	Ghana	56	Ghana	60
Guatemala	66	Guatemala	74	Guatemala	68
Guinea	17	Guinea	17	Guinea	18
Guyana	83	Guyana	79	Guyana	73

Haiti	41 Haiti	40 Haiti	31
Honduras	84 Honduras	87 Honduras	90
India	92 India	93 India	87
Indonesia	87 Indonesia	91 Indonesia	86
Iran, Islamic	93 Iran, Islamic	92 Iran, Islamic	83
Iraq	75 Iraq	75 Iraq	62
Cote d'Ivoire	48 Cote d'Ivoire	48 Cote d'Ivoire	42
Jamaica	86 Jamaica	87 Jamaica	74
Jordan	92 Jordan	92 Jordan	87
Kenya	74 Kenya	71 Kenya	59
Korea, DPR	98 Korea, DPR	99 Korea, DPR	99
Korea, Rep	74 Korea, Rep	74 Korea, Rep	95
Kuwait	94 Kuwait	94 Kuwait	98
Lao PDR	18 Lao PDR	26 Lao PDR	13
Lebanon	82 Lebanon	82 Lebanon	39
Lesotho	76 Lesotho	75 Lesotho	76
Liberia	28 Liberia	28 Liberia	55
Libya	84 Libya	84 Libya	70
Madagascar	46 Madagascar	46 Madagascar	33
Malawi	81 Malawi	79 Malawi	80
Malaysia	91 Malaysia	90 Malaysia	90
Mali	42 Mali	42 Mali	43
Mauritania	28 Mauritania	28 Mauritania	33
Mauritius	90 Mauritius	90 Mauritius	84
Mexico	66 Mexico	96 Mexico	78
Mongolia	84 Mongolia	85 Mongolia	86
Morocco	81 Morocco	81 Morocco	79
Mozambique	46 Mozambique	46 Mozambique	58
Oman	96 Oman	96 Oman	96
Namibia	53 Namibia	53 Namibia	41
Nepal	79 Nepal	78 Nepal	67
Nicaragua	65 Nicaragua	86 Nicaragua	82
Niger	13 Niger	13 Niger	21
Nigeria	57 Nigeria	57 Nigeria	54
Pakistan	96 Pakistan	96 Pakistan	97
Panama	82 Panama	82 Panama	99
Papua	69 Papua	69 Papua	67
Paraguay	78 Paraguay	76 Paraguay	69
Peru	72 Peru	73 Peru	64
Philippines	88 Philippines	88 Philippines	85
Guinea–Bis:	38 Guinea–Bis:	38 Guinea–Bis:	42
Qatar	82 Qatar	82 Qatar	79
Rwanda	84 Rwanda	83 Rwanda	83
Saudi Arabia	94 Saudi Arabia	94 Saudi Arabia	90
Senegal	60 Senegal	66 Senegal	59
Sierra Leone	83 Sierra Leone	83 Sierra Leone	75
Singapore	85 Singapore	85 Singapore	87
Viet Nam	87 Viet Nam	87 Viet Nam	87
Somalia	18 Somalia	18 Somalia	30
Zimbabwe	73 Zimbabwe	72 Zimbabwe	69
Sudan	62 Sudan	62 Sudan	57

Suriname	83	Suriname	81	Suriname	65
Swaziland	89	Swaziland	89	Swaziland	85
Syrian Arab	90	Syrian Arab	90	Syrian Arab	87
Thailand	92	Thailand	92	Thailand	80
Togo	61	Togo	61	Togo	57
Trinidad	83	Trinidad	83	Trinidad	69
U. Arab E.	85	U. Arab E.	85	U. Arab E.	75
Tunisia	90	Tunisia	90	Tunisia	87
Turkey	84	Turkey	84	Turkey	78
Uganda	77	Uganda	77	Uganda	74
Egypt	87	Egypt	87	Egypt	86
Tanzania	85	Tanzania	82	Tanzania	83
Burkina F.	37	Burkina F.	37	Burkina F.	42
Uruguay	88	Uruguay	88	Uruguay	82
Venezuela	63	Venezuela	72	Venezuela	62
Yemen AR	89	Yemen AR	89	Yemen AR	74
Zambia	79	Zambia	78	Zambia	76
WORLD	83	WORLD	85	WORLD	80
AFRICA	56	AFRICA	55	AFRICA	54
ASIA	90	ASIA	91	ASIA	88

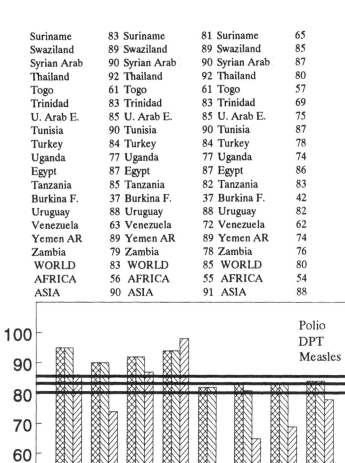

Figure 2.1. Percent immunized for selected countries; checkered pattern is for DPT, adjacent striped pattern is for polio, and third pattern (also striped) is for measles. Horizontal lines represent levels of World percentages for each of the three variables.

Although bar charts of this sort can be quite effective, they are always severely limited in size. With large data sets, another spatial approach that is related is to use maps and display the data on the maps. This can also be done electronically when the database with the WRD data is joined to the underlying database in a Geographic Information System (GIS) which can then be used to map the data.

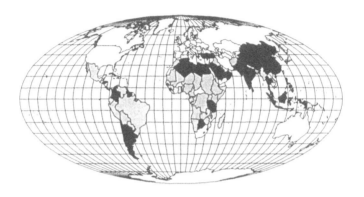

Figure 2.2. DPT immunization of 1-year-olds. World percentage is 83% of included nations. Included nations are shaded with light or dark shading. Those with light shading have immunization rates of less than 83%; those with dark shading have immunization rates greater than or equal to 83%. Mollweide equal area projection; scale, 1 inch represents 13,750 miles.

In Figure 2.2, the data from Table 2.2 has been joined to the database of a GIS (MapInfo for the PC). The map can then be colored to show which countries have immunization rates greater than, or less than, the World percentage. This strategy not only gives the researcher the capability to visualize large data sets in a bounded space, but it also can suggest associations of nations, as regions, that might have gone unnoticed in the table format of the data. The horizontal lines of Figure 2.1 have become the cut off lines for classes in the thematic maps of Figures 2.2, 2.3, and 2.4. Because each column of data in Table 2.2 (DPT, Polio, and Measles) is represented alone on a single map, national data can be compared to World data for each variable.

The careful reader will note that there is data for Burkina Faso in Table 2.2 but that the country has not been colored on the thematic map. The reason is that this country in western Africa is called "Burkina Faso" (its current name) in the WRD data base but is called "Upper Volta" (a previous name) in the MapInfo data base. Problems of this sort can usually be easily adjusted--simply by changing the name in one spreadsheet to match that in the other; what is important is to note them. The maps in Figures 2.2, 2.3, and 2.4 have been left with gaps of this sort in order to display this spatial data gap; in a presentation to policy makers, these gaps should be cleaned.

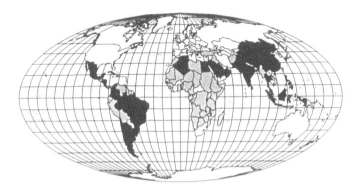

Figure 2.3. Polio immunization of 1-year-olds. World percentage is 85% of included nations. Included nations are shaded with light or dark shading. Those with light shading have immunization rates of less than 85%; those with dark shading have immunization rates greater than or equal to 85%. Mollweide equal area projection; scale, 1 inch represents 13,750 miles.

When the three maps are viewed in succession, it becomes clear that the region of the world that appears to have the worst immunization record, as a broad region, is Saharan and sub-Saharan Africa--a region that also has suffered devastating environmental degradation and destruction in recent years. The evidence of maps is often useful in cordoning off regions of adjacent countries with similar data characteristics that might not otherwise be noticed. Maps are two dimensional; tables are one dimensional (as a long string of data). The extra dimension in maps permits one to note territorial adjacencies that are not present in tabular data.

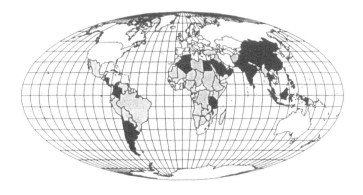

Figure 2.4. Measles immunization of 1-year-olds. World percentage is 80% of included nations. Included nations are shaded with light or dark shading. Those with light shading have immunization rates of less than 80%; those with dark shading have immunization rates greater than or equal to 80%. Mollweide equal area projection; scale, 1 inch represents 13,750 miles.

Because software interfaces often offer extra opportunity for the introduction of unusual and unexpected errors, one should also question visual displays in reports and presentations to understand what sorts of software linkages were involved in their formulation. The systematic questioning of how interface problems were recognized, and solved, is important in knowing whether or not the final map, chart, or figure is one that is based on a consistent data set.

Consistent database construction by estimation
The preceding analysis was derived from eliminating the TB variable from Table 2.1 in order to generate a consistent data set. Instead, one might keep the TB variable and eliminate, instead, countries from the DPT, Polio, and Measles data sets that do not also appear in the TB data set. These countries are: Bahrain, Cyprus, Jordan, Kuwait, Lebanon, Suriname, Trinidad (and Tobago), and Turkey. They were the countries graphed in Figure 2.1; lack of data overlap can serve as a reason to group countries. That figure suggests that a number of these excluded countries have fine immunization programs on the variables for which there is data.

Because immunization programs may be constructed to offer a range of vaccinations, rather than just a single one, one might consider that a country with high percentage rates in immunizations on three of four variables would have a high immunization rate on the fourth (if the disease is indigenous). If such a relationship exists (it need not), it would then become possible to estimate the missing values on the basis of the values that are present. Information can be estimated when there is substantial actual reason (as might be the case in immunization programs) for suspecting a causal relationship between variables; this sort of procedure can produce misleading results if not done with care.

In Table 2.3, the eight countries that do not appear in all four columns of Table 2.1 have been removed, and the columns adjusted so that they are all the same length. Then, each column of country name and percentage was sorted numerically, in ascending order. The continental and global entries were sorted, too. Thus, one can see simply by running an eye down each column, which African countries are above or below the percentage for all of Africa. These continental and global entries were then removed. The resulting table (almost identical to Table 2.3) was used to find regression equations, as follows.

STRAIGHT LINE FIT TO THE DATA OF TABLE 2.3
see Tables 2.3 and 2.4

1. Enter the values for DPT (as in Table 2.3--national data only), the independent variable, in spreadsheet column (A) to be used as the x-axis input.
2. Enter the values for TB (as in Table 2.3--national data only), the dependent variable, in another column (spreadsheet column B), to be used as y values
3. Choose the regression feature from the software, with the x values as in step 1 and the y values as in step 2.
4. Choose the output range as a blank area in the spreadsheet. Then proceed with the calculation as directed by the software; the output from the regression will appear in a form similar to the one in Table 2.4, upper left (produced in Lotus 1-2-3, release. 2.3).
5. The equation below the output range must generally be derived by the user from the regression output. The slope-intercept form for the equation of a straight line ($y=mx+b$) is used. The "X Coefficient" from the regression output is used as "m". The "Constant" from the regression output is used as b. This equation is determined from actual data.

TABLE 2.3 (source: World Resources Institute)

Immunization of 1−year−olds   Percent, 1990.

	TB	DPT	Polio	Measles
Lao PDR	26 Niger	13 Niger	13 Lao PDR	13
Afghanistan	30 Guinea	17 Guinea	17 Guinea	18
Somalia	31 Lao PDR	18 Somalia	18 Afghanistan	20
Angola	47 Somalia	18 Chad	20 Niger	21
Bolivia	48 Chad	20 Angola	23 Somalia	30
Niger	50 Angola	23 Afghanistan	25 Haiti	31
Guinea	53 Afghanistan	25 Lao PDR	26 Zaire	31
Cambodia	54 Liberia	28 Liberia	28 Chad	32
Ethiopia	56.96 Mauritania	28 Mauritania	28 Madagascar	33
Chad	59 Zaire	32 Zaire	31 Mauritania	33
Mozambique	59 Burkina F.	37 Burkina F.	37 Cambodia	34
El Salvador	60 Guinea−Bis:	38 Guinea−Bis:	38 Ethiopia	37.12
Guatemala	62 Cambodia	40 Cambodia	40 Angola	38
Liberia	62 Bolivia	41 Haiti	40 Namibia	41
Cote d'Ivoire	63 Haiti	41 Mali	42 Burkina F.	42
Venezuela	63 Mali	42 Ethiopia	44.16 Cote d'Ivoire	42
Zaire	65 Ethiopia	44.16 Madagascar	46 Guinea−Bis:	42
Madagascar	67 Madagascar	46 Mozambique	46 Mali	43
Dom. Rep.	68 Mozambique	46 Cote d'Ivoire	48 Bolivia	53
Mexico	70 Cote d'Ivoire	48 Bolivia	50 AFRICA	54
Honduras	71 Namibia	53 Namibia	53 Bangladesh	54
Zimbabwe	71 AFRICA	56 Cameroon	54 Nigeria	54
Haiti	72 Cameroon	56 AFRICA	55 Liberia	55
Korea, Rep	72 Ghana	57 Ghana	56 Cameroon	56
Sudan	73 Nigeria	57 Nigeria	57 Sudan	57
Mauritania	75 Senegal	60 Togo	61 Togo	57
Myanmar	75 Togo	61 Bangladesh	62 Mozambique	58
Cameroon	76 Bangladesh	62 Sudan	62 Kenya	59
Brazil	78 Sudan	62 Senegal	66 Senegal	59
AFRICA	80 Venezuela	63 Benin	67 Ghana	60
Belize	80 Nicaragua	65 Ecuador	67 Ecuador	61
Kenya	80 Guatemala	66 Myanmar	69 Iraq	62
Ghana	81 Mexico	66 Papua	69 Venezuela	62
Nicaragua	81 Benin	67 Kenya	71 Peru	64
Mali	82 Ecuador	68 Venezuela	72 Nepal	67
Panama	83 Dom. Rep.	69 Zimbabwe	72 Papua	67
Peru	83 Myanmar	69 Peru	73 Guatemala	68
Burkina F.	84 Papua	69 Guatemala	74 Paraguay	69
Guyana	85 Peru	72 Korea, Rep	74 Zimbabwe	69
Namibia	85 Zimbabwe	73 Eq. Guinea	75 Benin	70
Bangladesh	86 Kenya	74 Iraq	75 Libya	70
Solomon Is.	87 Korea, Rep	74 Lesotho	75 Solomon Is.	70

Ecuador		Iraq	88	Solomon Is.	75	Gambia	73
Egypt		El Salvador	88	El Salvador	76	Guyana	73
Sri Lanka		Lesotho	88	Paraguay	76	Myanmar	73
Papua		Solomon Is.	89	Uganda	77	Jamaica	74
WORLD		Uganda	90	Gabon	77	Uganda	74
Congo		Eq. Guinea	90	Nepal	78	Yemen AR	74
Guinea–Bis:		Gabon	90	Zambia	78	Burundi	75
Libya		Paraguay	90	Congo	78	Congo	75
Paraguay		Congo	90	Guyana	79	El Salvador	75
Viet Nam		Nepal	90	Malawi	79	Sierra Leone	75
Benin		Zambia	92	Belize	79	U. Arab E.	75
Botswana		Brazil	92	Morocco	81	Gabon	76
Costa Rica		Malawi	92	Botswana	81	Lesotho	76
Mongolia		Morocco	92	C. Afr. Rep.	81	Zambia	76
Rwanda		C. Afr. Rep.	92	Panama	82	Botswana	78
Senegal		Panama	92	Qatar	82	Brazil	78
Syrian Arab :		Qatar	92	Tanzania	82	Mexico	78
Indonesia		WORLD	93	Rwanda	83	Cape Verde	79
Oman		Guyana	93	Sierra Leone	83	Morocco	79
Tanzania		Sierra Leone	93	Libya	83	Qatar	79
Mauritius		Belize	94	WORLD	84	WORLD	80
Togo		Honduras	94	Djibouti	84	Malawi	80
ASIA		Libya	95	Mongolia	84	Thailand	80
Barbados		Mongolia	95	Singapore	84	Belize	81
Colombia		Rwanda	95	U. Arab E.	84	Colombia	82
Djibouti		Argentina	95	Burundi	85	C. Afr. Rep.	82
Iran, Islamic		Djibouti	95	Nicaragua	85	Nicaragua	82
C. Afr. Rep.		Singapore	96	Cape Verde	85	Uruguay	82
Gabon		Tanzania	96	Egypt	85	Algeria	83
Iraq		U. Arab E.	96	Honduras	85	Iran, Islamic	83
Morocco		Botswana	96	Jamaica	86	Rwanda	83
Nigeria		Burundi	96	Viet Nam	86	Sri Lanka	83
Philippines		Jamaica	96	Philippines	86	Tanzania	83
Swaziland		Colombia	96	Uruguay	87	Fiji	84
U. Arab E.		Egypt	96	Algeria	87	Mauritius	84
Burundi		Indonesia	97	Argentina	87	Djibouti	85
Cape Verde		Viet Nam	97	Swaziland	87	Philippines	85
Chile		Cape Verde	97	Yemen AR	88	Swaziland	85
Eq. Guinea		Philippines	97	Barbados	88	Egypt	86
India		Uruguay	97	Dom. Rep.	88	Indonesia	86
Lesotho		Algeria	97	Malaysia	89	Mongolia	86
Malawi		Swaziland	97	Mauritius	89	Barbados	87
Nepal		Yemen AR	97	Sri Lanka	89	Comoros	87
Qatar		ASIA	97	Syrian Arab :	90	India	87
Zambia		Gambia	97	Tunisia	90	Singapore	87
Cuba		Mauritius	98	ASIA	90	Syrian Arab :	87
Jamaica		Sri Lanka	98	Indonesia	90	Tunisia	87
Pakistan		Syrian Arab :	98	Iran, Islamic	90	Viet Nam	87
Sierra Leone		Tunisia	98	Thailand	90	ASIA	88
Algeria		Barbados	99	Brazil	91	Eq. Guinea	88
Argentina		Malaysia	99	Colombia	91	Bhutan	89
Bhutan		Cuba	99	Gambia	92	Costa Rica	90

China	99	India	92	India	93	Honduras	90
Comoros	99	Thailand	92	Comoros	94	Malaysia	90
Fiji	99	Iran, Islamic	93	Cuba	94	Saudi Arabia	90
Gambia	99	Comoros	94	Saudi Arabia	94	Cuba	94
Korea, DPR	99	Saudi Arabia	94	Bhutan	95	Argentina	95
Malaysia	99	Bhutan	95	Costa Rica	95	Korea, Rep	95
Saudi Arabia	99	Costa Rica	95	Fiji	96	Dom. Rep.	96
Singapore	99	Oman	96	Mexico	96	Oman	96
Thailand	99	Pakistan	96	Oman	96	Pakistan	97
Tunisia	99	China	97	Pakistan	96	Chile	98
Uganda	99	Fiji	97	China	98	China	98
Uruguay	99	Korea, DPR	98	Chile	99	Korea, DPR	99
Yemen AR	99	Chile	99	Korea, DPR	99	Panama	99

6. The linear equation for this case, y=0.754502*x+30.13463, may be used to find TB values for the remaining eight countries, derived from DPT immunization information.

This particular regression used DPT as an independent variable to attempt to predict values for TB. The correlation was made on actual data; the fit was quite good--R-squared is 0.97 (a perfect fit has an R-squared value of 1.0). When Polio and Measles are used as independent variables, regression analysis may also be performed, obtaining an equation for a line fit to actual values which may then be extended to estimate values for the TB variable for the eight countries for which there is no information. The procedure is given in detail below.

STRAIGHT LINE FIT TO THE DATA OF TABLE 2.3
see Tables 2.3 and 2.4

1. Enter the values for Polio (as in Table 2.3--national data only), the independent variable, in spreadsheet column (A) to be used as the x-axis input.
2. Enter the values for TB (as in Table 2.3--national data only), the dependent variable, in another column (spreadsheet column B), to be used as y values
3. Choose the regression feature from the software, with the x values as in step 1 and the y values as in step 2.

TABLE 2.4

All entries, TB = f(DPT)
Regression Output:

Constant	30.13463
Std Err of Y Est	2.961430
R Squared	0.970386
No. of Observations	104
Degrees of Freedom	102
X Coefficient(s)	0.754502
Std Err of Coef.	0.013050

Linear equation:
y=0.754502*x+30.13463

All but 3 entries, TB = f(DPT)
Regression Output:

Constant	34.51538
Std Err of Y Est	1.528050
R Squared	0.988867
No. of Observations	101
Degrees of Freedom	99
X Coefficient(s)	0.700147
Std Err of Coef.	0.007466

Linear equation:
y=0.700147*x+34.51538

All entries, TB = f(Polio)
Regression Output:

Constant	29.49125
Std Err of Y Est	2.893531
R Squared	0.971729
No. of Observations	104
Degrees of Freedom	102
X Coefficient(s)	0.752477
Std Err of Coef.	0.012708

Linear equation:
y=0.752477*x+29.49125

All but 3, TB = f(Polio)
Regression Output:

Constant	33.77739
Std Err of Y Est	1.596643
R Squared	0.987845
No. of Observations	101
Degrees of Freedom	99
X Coefficient(s)	0.699911
Std Err of Coef.	0.007802

Linear equation:
y=0.699911*x+33.77739

All entries, TB = f(Measles)
Regression Output:

Constant	28.06999
Std Err of Y Est	3.541376
R Squared	0.957652
No. of Observations	104
Degrees of Freedom	102
X Coefficient(s)	0.800902
Std Err of Coef.	0.016675

Linear equation:
y=0.800902*x+28.06999

All but 3, TB = f(Measles)
Regression Output:

Constant	32.55265
Std Err of Y Est	2.587397
R Squared	0.968081
No. of Observations	101
Degrees of Freedom	99
X Coefficient(s)	0.743657
Std Err of Coef.	0.013571

Linear equation:
y=0.743657*x+32.55265

4. Choose the output range as a blank area in the spreadsheet. Then proceed with the calculation as directed by the software; the output from the regression will appear in a form similar to the one in Table 2.4, middle left (produced in Lotus 1-2-3, release. 2.3).

5. The equation below the output range must generally be derived by the user from the regression output. The slope-intercept form for the equation of a straight line (y=mx+b) is used. The "X Coefficient" from the regression output is used as "m". The "Constant" from the regression output is used as b. This equation is determined from actual data.

6. The linear equation for this case, y=0.752477*x+29.49125, may be used to find TB values for the remaining eight countries, derived from Polio immunization information.

This particular regression used Polio as an independent variable to attempt to predict values for TB. The correlation was made on actual data; the fit was quite good--R-squared is 0.97 (a perfect fit has an R-squared value of 1.0).

STRAIGHT LINE FIT TO THE DATA OF TABLE 2.3
see Tables 2.3 and 2.4.

1. Enter the values for Measles (as in Table 2.3--national data only), the independent variable, in spreadsheet column (A) to be used as the x-axis input.

2. Enter the values for TB (as in Table 2.3--national data only), the dependent variable, in another column (spreadsheet column B), to be used as y values

3. Choose the regression feature from the software, with the x values as in step 1 and the y values as in step 2.

4. Choose the output range as a blank area in the spreadsheet. Then proceed with the calculation as directed by the software; the output from the regression will appear in a form similar to the one in Table 2.4, lower left (produced in Lotus 1-2-3, release. 2.3).

5. The equation below the output range must generally be derived by the user from the regression output. The slope-intercept form for the equation of a straight line (y=mx+b) is used. The "X Coefficient" from the regression output is used as "m". The "Constant" from the regression output is used as b. This equation is determined from actual data.

6. The linear equation for this case, y=0.800902*x+29.49125; it may be used to find TB values for the remaining eight countries, derived from Measles immunization information.

TABLE 2.5

Regression estimates of TB based on DPT, Polio, Measles.

	DPT	1 RMS−	1 RMS+	2 RMS−	2 RMS+	TB estimate
Bahrain	95	99.50129	102.5	97.97324	104.0	101.0293
Cyprus	90	96.00056	99.05	94.47251	100.5	97.52861
Jordan	92	97.40085	100.4	95.87280	101.9	98.92890
Kuwait	94	98.80114	101.8	97.27309	103.3	100.3291
Lebanon	82	90.39938	93.45	88.87133	94.98	91.92743
Suriname	83	91.09953	94.15	89.57148	95.68	92.62758
Trinidad	83	91.09953	94.15	89.57148	95.68	92.62758
Turkey	84	91.79967	94.85	90.27162	96.38	93.32772
	Polio					
Bahrain	95	98.67229	101.8	97.07564	103.4	98.67229
Cyprus	90	95.17273	98.36	93.57609	99.96	95.17273
Jordan	92	96.57255	99.76	94.97591	101.3	96.57255
Kuwait	94	97.97238	101.1	96.37573	102.7	97.97238
Lebanon	82	89.57344	92.76	87.97680	94.36	89.57344
Suriname	81	88.87353	92.06	87.27689	93.66	88.87353
Trinidad	83	90.27336	93.46	88.67671	95.06	90.27336
Turkey	84	90.97327	94.16	89.37662	95.76	90.97327
	Measles					
Bahrain	86	93.91975	99.09	91.33235	101.6	93.91975
Cyprus	74	84.99587	90.17	82.40847	92.75	84.99587
Jordan	87	94.66341	99.83	92.07601	102.4	94.66341
Kuwait	98	102.8436	108.0	100.2562	110.6	102.8436
Lebanon	39	58.96787	64.14	56.38047	66.73	58.96787
Suriname	65	78.30295	83.47	75.71556	86.06	78.30295
Trinidad	69	81.27758	86.45	78.69018	89.03	81.27758
Turkey	78	87.97049	93.14	85.38310	95.73	87.97049

This particular regression used Measles as an independent variable to attempt to predict values for TB. The correlation was made on actual data; the fit was quite good--R-squared is 0.95 (Table 2.4--a perfect fit has an R-squared value of 1.0).

BLACK BOX SUMMARY
see Introduction for theoretical explanation

LEAST SQUARES REGRESSION LINE

y=mx+b

where
m is the slope of the line, or the "x-coefficient"
b is the y-intercept, or the "constant."

The first three values for each variable in Table 2.3 are relatively far from the regression line; thus, we can improve the R-squared value by removing these three low values, which will not be used as DPT, Polio, or Measles inputs for the eight countries. Then, repeat the regression analyses going through the same steps as above, but with these truncated columns of Table 2.3 data. When these procedures are executed, the R-squared values are as on the right side of Table 2.4, and the corresponding regression equations, y=0.700147*x+34.51538 for DPT with an R-squared of 0.988, y=0.699911*x+33.77739 for Polio with an R-squared of 0.987, and y=0.743657*x+32.55265 for Measles with an R-squared of 0.968, are given on the right side of Table 2.4. These improved estimating equations are then used to generate estimated values for TB, as a dependent variable, based on each of DPT, Polio, and Measles as input for the independent variable.

The estimates, using the improved DPT regression equation for each of the eight countries are: 101.0293, 97.52861, 98.92890, 100.3291, 91.92743, 92.62758, 92.62758, and 93.32722, as shown in Table 2.5.

The estimates, using the improved Polio regression equation for each of the eight countries are: 98.67229, 95.17273, 96.57255, 97.97238, 89.57344, 88.87353, 90.27336, and 90.97327, as shown in Table 2.5.

The estimates, using the improved Measles regression equation for each of the eight countries are: 93.91975, 84.99587, 94.66341, 102.8436, 58.96787, 78.30295, 81.27758, and 87.97049, as shown in Table 2.5.

Some of the estimated values for these projections are greater than 100; while that does not make sense in terms of percentage, it is simply the numerical value that is produced using x=95 (for example) as the input for Bahrain, in the regression equation; that substitution generates an output of 101.0293. The other values in Table 2.5--the RMS columns-- show the Root Mean Square, derived from subtracting and adding the standard error of the y estimates (1.528050 for example, see Table 2.4) from the regression equation (hence 1 RMS- and 1 RMS+). When twice the standard y-error is subtracted from and added to the regression equation, the values in the columns labelled 2RMS- and 2RMS+ are obtained. The standard error creates a strip, or buffer, on either side of the regression line; generally, one would expect that about 68% of the time, the y estimate (for arbitrary x) would fall between the limits of lower and upper 1RMS. Further, one would expect that about 95% of the time, in a linear scatter of dots, the regression estimate would fall between twice the RMS. This measure gives the user an assessment of the level of confidence in the estimate; whether or not the estimate with the highest R-squared value, or smallest y-error, is chosen might well depend on conditions that are more than numerical--such as whether or not one disease is more likely in a particular region than in another. Cultural constraints often circumscribe numerical decisions; but once some set of estimated values for the missing entries is included, then Table 2.1 can be rewritten in a consistent manner, and a database for all four variables has been constructed in a defensible manner.

Commentary

Immunization is but one view of disease control. Some low income countries commit so much of their agricultural resource base to non-food production, or tropical food exports that the nutrition of their population suffers--fats and oils and cereals are more often used for domestic consumption. Beverages and non-food agricultural products are for export. Thus, one way to measure health/disease control might be to correlate percent child malnutrition, through stunting/wasting, with commodity indices involving, non-food agricultural production, total agriculture, total food, beverages, and fats and cereals. Extra sources of data would be required; the two databases employed here have at most global data only on the commodity indices. Lack of appropriate data can often inhibit investigation of an otherwise interesting and important theme, underscoring the significance of employing creatively those techniques that are available.

CHAPTER 3

AGRICULTURE DATA ANALYSIS

ANALYTICAL TECHNIQUES/TOOLS USED

Linear fitting--bounded
Bar charts
Default computer curves
Linear fitting--unbounded; simple use of least squares analysis
Exponential fitting--unbounded
Extrapolation of fitted curves: prediction

DATA TYPE: ABUNDANT LONGITUDINAL DATA

Overview of Data

Various international organizations publish data on agriculture and food in an electronic format; Table 3.1 displays data from one set (World Bank). With any data set (presented in electronic or paper format), it is important first to examine the set for interesting or unusual patterns in the display. These patterns often influence decisions in choosing subsets of data and tools to analyze subsets.

PATTERNS IN DATA--WHAT TO LOOK FOR

1. What is the general organizational scheme of the entire set? Is it arranged alphabetically, numerically, or in some other fashion?

2. Are the real-world entries in the Table (nations, states, counties) expressed as comparable units? For example, county data and national data are generally not comparable.

3. Are the numerical entries in the Table expressed in comparable units? For example, data in one column might measure percentages while data in another column might measure thousands of dollars--these columns would not be comparable.

4.   Are there gaps in the data? If so, what is their significance to
the questions you wish to have the data answer?

Using these four items as a guide, consider the data set displayed in
Table 3.1.

1.   Countries are organized roughly from low-income economies to
high-income economies. Each is designated by a three letter code
roughly corresponding to its name, or in some cases, to an earlier name.
Thus, Myanmar has a three letter code BUR for Burma, its old name.
Burkina Faso is shortened to HVO, for the French "Haute Volta" or
"Upper Volta," a previous name. Each three letter designation is
unique; the entire set should be considered prior to assuming what a
particular three letter code might represent. Lack of care in this
regard might lead a reader of this Table (or any other using this form of
code) to confuse, for example, Zambia, ZMB, with Zimbabwe, ZWE.

2.   Not all entries represent data for a single country; some are
groupings of nations--as for example, "Oil exporters." Analytical
techniques should therefore not be applied alike to single nations and to
groupings of nations.

3. The numbers in the table represent different units; thus, one should
not assume that a value of 93 in the first column and a value of 93 in
the last column represent the same amounts. Indeed, a value of 93 in the
first column is expressed in millions of current U.S. dollars; a value of
93 in the last column is a percentage. Prior to analysis, units should be
converted to a format (if necessary) that will produce meaningful
results.

4. There are a number of blank entries grouped in various locations in
the columns.   One should consider why these are there, prior to
choosing a subset of data or an analytical technique.   In particular,
toward the end of the columns that give values for   "food aid in
cereals," there are many single countries that show no entries, either in
1974 or in 1989. Presumably, these are nations that can supply their
needs with what they raise or with what they buy as imports from other
nations. They are wealthier and do not need aid. Thus they are not
comparable, in this regard, to low income countries receiving aid.

TABLE 3.1 (source: World Bank)
AGRICULTURE AND FOOD
Column labels at end of table.

	COL. 01	COL. 02	COL. 03	COL. 04	COL. 05	COL. 06	COL. 07	COL. 08	COL. 09	
G01	89,314	305,959	22,608	28,763	6,002	5,235	171	802	116	Low–income
G02	60,621	205,278	11,294	15,014	1,582	531	241	1,185	122	China and India
G03	28,269	99,716	11,314	13,749	4,420	4,704	72	310	103	Other low–income
MOZ		704	62	400	34	424	22	21	83	Mozambique
ETH	931	2,254	118	690	54	573	4	39	89	Ethopia
TZA	473	1,795	431	83	148	76	31	92	90	Tanzania
SOM	167	705	42	186	111	73	29	40	97	Somalia
BGD	3,636	8,962	1,866	2,204	2,076	1,161	157	770	93	Bangladesh
LAO			53	64	8	20	2	6	116	Lao
MWI	119	498	17	86	0	217	52	203	85	Malawi
NPL	579	1,633	18	26	0	9	27	232	107	Nepal
TCD	142	364	37	37	20	15	7	17	101	Chad
BDI	159	535	7	6	6	6	5	20	98	Burundi
SLE	108	409	72	145	10	38	17	3	89	Sierra Leone
MDG	243	717	114	103	7	76	61	21	93	Madagascar
NGA	5,080	8,874	389	240	7	0	2	94	96	Nigeria
UGA	929	2,986	36	16		17	14	2	87	Uganda
ZAR	805	2,846	343	323	1	55	8	9	94	Zaire
MLI	207	1,048	281	89	107	62	31	59	97	Mali
NER	420	744	155	105	73	83	1	8	86	Niger
HVO	121	871	99	120	28	49	3	57	115	Burkina Faso
RWA	135	799	3	10	19	2	3	20	77	Rwanda
IND	23,916	71,345	5,261	1,014	1,582	308	137	517	113	India
CHN	36,705	133,934	6,033	14,000	0	223	410	2,361	128	China
HTI		692	83	251	25	49	4	25	93	Haiti
KEN	484	2,208	15	119	2	112	238	421	101	Kenya
PAK	3,352	9,681	1,274	2,171	584	416	146	829	103	Pakistan
BEN	121	729	7	104	9	16	36	49	114	Benin
CAF	60	442	7	28	1	0	12	4	90	Cen. African Rep.
GHA	1,030	2,570	177	244	33	46	13	38	109	Ghana
TGO	85	446	6	111	11	11	3	76	89	Togo
ZMB	191	617	93	123	5	66	73	183	97	Zambia
GIN		812	63	183	49	42	19	6	90	Guinea
LKA	545	1,648	951	1,177	271	272	555	1,094	87	Sri Lanka
LSO	23	83	48	140	14	34	10	125	80	Lesotho
IDN	4,340	22,032	1,919	2,356	301	69	133	1,068	124	Indonesia
MRT	58	339	115	207	48	70	11	55	88	Mauritania
AFG			5	260	10	208	24	97		Afghanistan
BTN		125	3	20	0	2		10	121	Bhutan
KHM			223	50	226	11	11	2		Kampuchea
LBR	91		42	158	3	28	63	94	95	Liberia
BUR			26		9	0	21	125	120	Myanmar
SDN	757		125	556	46	198	28	40	87	Sudan
VNM			1,854	258	64	100	513	651	111	Viet Nam
G04	50,052	258,932	42,817	80,767	1,926	4,548	370	703	101	Middle income
G05	31,154	129,238	24,693	49,426	1,654	4,510	309	592	99	Lower middle income
AGO			149	248	0	79	33	29	84	Angola
BOL	202	1,440	209	172	22	95	7	19	102	Bolivia
EGY	1,942	5,858	3,877	8,543	610	1,427	1,312	3,505	109	Egypt
SEN	208	1,028	341	515	27	53	17	40	106	Senegal
YEM			306	1,378	33	85		63		Yemen
ZWE	214	664	56	52	0	10	446	505	90	Zimbabwe
PHL	1,996	10,429	817	1,626	89	135	287	612	86	Philippines
CIV	462	3,295	172	693	4	19	74	90	96	Cote d'Ivoire
DOM	345	1,012	252	601	16	228	334	556	94	Dominican Rep.
MAR	789	3,679	891	1,329	75	238	117	376	120	Morocco
PNG	240	1,000	71	243		0	58	381	97	Papua New Guinea
HND	212	890	52	172	31	67	156	190	88	Honduras
GTM		1,472	138	214	9	277	298	656	103	Guatemala
COG	49	311	34	82	2	2	114	25	98	Congo, PR
SYR	435	2,475	339	1,578	47	31	68	404	86	Syria
CMR	364	2,978	81	345	4	6	34	71	96	Cameroon
PER	1,351	2,177	637	1,065	37	146	300	622	101	Peru
ECU	401	1,526	152	536	13	89	133	232	106	Ecuador
NAM		187		0					95	Namibia
PRY	191	1,217	71	5	10	1	98	69	115	Paraguay

Code									Index	Country
SLV	292	685	75	186	4	197	1,043	1,262	90	El Salvador
COL	1,806	6,622	502	716	28	12	287	945	102	Colombia
THA	1,837	10,561	97	346	0	83	59	328	104	Thailand
JAM	93	210	340	296	1	365	873	914	92	Jamaica
TUN	245	1,235	307	1,655	59	284	76	222	96	Tunisia
TUR	3,383	11,857	1,276	3,061	16	3	157	637	97	Turkey
BWA	28	75	21	77	5	33	15	7	68	Botswana
JOR		241	171	671	79	25	74	362	117	Jordan
PAN	149	493	63	109	3		387	657	92	Panama
CHL	558		1,737	178	323	14	313	544	107	Chile
CRI	222	897	110	357	1	84	1,001	1,806	89	Costa Rica
POL			4,185	2,893			1,678	2,223	106	Poland
MUS	30	222	160	209	22	21	2,095	3,075	100	Mauritius
MEX	4,462	18,050	2,881	7,050		291	232	753	98	Mexico
ARG	2,250	7,339	0	4			26	45	91	Argentina
MYS	1,198		1,023	2,299	1	10	489	1,596	142	Malaysia
DZA	492	6,187	1,816	7,461	54	39	163	320	97	Algeria
BGR		1,758	649	1,384			1,411	1,804	100	Bulgaria
LBN	136		354	558	26	32	1,354	671		Lebanon
MNG			28	59			22	184	91	Mongolia
NIC	199		44	140	3	32	215	433	63	Nicaragua
G06	19,594		18,124	31,341	271	38	465	865	103	Upper middle income
VEN	826	2,654	1,270	1,804			170	1,580	88	Venezuela
ZAF	1,362	4,635	127	296			422	541	90	South Africa
BRA	4,392	27,810	2,485	2,015	31	15	186	485	115	Brazil
HUN	1,010	4,048	408	249			1,497	2,595	113	Hungary
URY	268	773	70	81	6		485	420	106	Uruguay
YUG	2,212	7,229	992	192			770	1,328	98	Yugoslavia
GAB	60	353	24	50				46	81	Gabon
IRN	2,120	34,563	2,076	6,500		23	60	658	87	Iran
TTO	40	118	208	284			880	450	86	Trinidad & Tobago
CSK		3,266	1,296	216			2,404	3,031	121	Czechoslovakia
PRT			1,861	1,226			428	1,026	100	Portugal
KOR	2,311	21,663	2,679	10,267	234		2,450	3,920	96	Korea
OMN	40	202	52	200				417		Oman
LBY	93		612	1,515			62	416	109	Libya
GRC	1,569		1,341	465			861	1,542	100	Greece
IRQ	579		870	4,891			34	397	98	Iraq
ROM			1,381	556			565	1,301	109	Romania
G07	141,602	571,792	65,426	109,529	7,928	9,783	256	758	112	Low/middle income
G08	15,597	52,090	4,208	7,411	910	2,610	34	76	95	Sub-Saharan Afr.
G09	49,792	211,600	14,938	31,795	923	651	365	1,712	123	East Asia
G10	32,884	103,077	9,404	6,634	4,522	2,169	135	541	112	South Asia
G11	20,496	116,812	25,193	46,909	1,010	2,394	575	1,058	99	Europe, M.East
G12	18,661		11,556	16,484	563	1,960	177	464	105	Latin Am.
G13	23,513		20,373	29,501	1,274	2,705	351	647	105	Severly indebted
G14	82,405		78,976	75,503			1,032	1,238	99	High–income
G15	80,527		72,941	64,224			1,029	1,221	98	OECD
G16	1,880	18,155	6,035	11,279			1,423	3,445	123	Other
SAU	219	6,150	482	5,560			54	3,678		Saudi Arabia
IRL	559	3,307	640	379			3,067	6,815	105	Ireland
ESP		18,160	4,675	2,224			593	989	111	Spain
ISR	295		1,176	1,890	53	4	1,401	2,237	106	Israel
HKG	62	184	657	826					61	Hong Kong
SGP	44	97	682	925			2,500	18,333	86	Singapore
NZL	897		92	190			7,745	7,086	107	New Zealand
AUS	2,157	10,402	2	26			232	286	96	Australia
GBR	2,993		7,540	2,908			2,631	3,555	105	Great Britain
ITA	8,365	30,579	8,101	7,649			896	1,901	100	Italy
NLD	1,827	9,155	7,199	5,932			7,493	6,877	110	Netherlands
KWT	8	238	101	597				750		Kuwait
BEL	920	3,165	4,585	4,004			5,648	5,098	116	Belgium
AUT	992	4,042	164	81			2,426	2,214	109	Austria
FRA	1,221	31,843	654	917			2,435	2,990	105	France
ARE		481	132	596				1,632		United Arab Emir.
CAN	3,265		1,513	1,067			191	484	103	Canada
DEU	5,951	18,307	7,164	4,524			4,263	4,208	112	Germany
DNK	882	3,942	462	171			2,234	2,330	120	Denmark
USA	27,812		460	2,147			816	937	92	U.S.A.
SWE		4,879	300	120			1,646	1,357	94	Sweden

FIN	1,205	5,808	222	214			1,930	2,164	102	Finland
NOR	624	2,757	713	545			2,443	2,704	109	Norway
JPN	12,467	72,773	19,557	27,370			3,547	4,327	97	Japan
CHE			1,458	651			3,831	4,306	102	Switzerland
G17			10,533	41,874		1	464	1,209	110	Other economies
G18	249,704		154,934	226,907	7,981	9,787	497	954	110	World
G19	9,822	65,457	8,166	29,579	63	143	49	408	101	Oil exporters
OAN										Taiwan

COLUMN LABELS   TABLE OF AGRICULTURE AND FOOD

01   Value added in agriculture (mill. current $) 1970
02   Value added in agriculture (mill. current $) 1989
03   Cereal imports (thous. metric tons) 1974
04   Cereal imports (thous. metric tons) 1989
05   Food aid in cereals (thous. metric tons) 1974/75
06   Food aid in cereals (thous. metric tons) 1988/89
07   Fert. cons.(100s gr.of plant nutrient/hectare) 1970/71
08   Fert cons. (100s gr.of plant nutrient/hectare) 1987/88
09   Avg. index of food prod. p.c. (1979–81=100) 1987–89

Maps used in conjunction with tables of data often suggest spatial patterns in the data which might otherwise go unnoticed. In Table 3.1, a particularly large number of the countries in the lowest income grouping are in Africa (Figure 3.1). It might therefore be natural to consider cereal imports, cereal aid, or any of the other indicators in Table 3.1 for Africa alone. With the data from Table 3.1 entered in an electronic spreadsheet, and an atlas, it is a relatively easy matter to sort Table 3.1 by continent. If the user of the data did not know which countries were in Africa, this association would not be apparent in the table. Some readers may already have good maps in their brains; others should make substantial use of maps and atlases to look for spatial pattern. Table 3.2 shows the subset of African data, arranged as in Table 3.1.

Linear Fitting: Some Bounded Curves

Curves that are created in an electronic environment are often formed from straight line segments linking data points, plotted in an x-y coordinate system. Thus, these electonically-produced curves are bounded--they do not extend beyond data points.

Most commonly available spreadsheets and other graphing software can create bar charts, curves, pie charts, and various other visual displays of data. Online help is usually sufficient to guide the computer-literate individual through a sequence of steps to make any of the graphs.

With reference to the data in Table 3.1, there are a number of ways one might graphically portray imports in cereal by country. Prior to doing so, however, it is important, for the following reasons, to consider whether the data being dealt with is discrete or continuous.

1. The countries are discrete entities; bar charts are often useful for displaying such data.

2. The values for imports, when change is focused upon, trace out a continuous set of values. Continuous curves are useful for displaying this sort of data.

What we have in this situation is a discrete variable (countries of Africa) plotted against a continuous one (change in imports). When the technique chosen is one suited to discrete data, the continuous is ignored (and vice versa). In short, there is a tradeoff, and users of software need to understand this sort of distinction. When the basic nature of the tool matches the basic nature of the data, all works well; typically, however, this is not the case, and then one needs to consider carefully what sorts of decisions are involved in getting a good graphic portrayal of the data.

The data in Table 3.1 concerning cereal imports are given in thousands of metric tons. Thus, in 1974 Mozambique (MOZ) imported 62 thousand metric tons of cereal. If the number 62 is entered as a value, the thousand that is understood by the reader is lost in the electronic environment. Thus, some sort of conversion, so that electronically-generated labels are correct, is needed. It is an easy matter, in this case, to multiply each entry by 1000. In Table 3.2, the entries in the columns of "cereal import" data have been multiplied by 1000.

Bar charts
When the country name is the variable on the x-axis and millions of metric tons of cereal imports is the variable on the y-axis, a bar chart showing data from 1974 and 1989, for 43 countries, is easy to construct. In Figure 3.2, contiguous vertical bars represent data for a single country for each of 1974 and 1989. This sort of display has the advantage of exhibiting direct comparisons for each country. No clear picture of the broad sweep of change in imports emerges--either for a single year, or for the two years relative to each other. The discrete is emphasized; the continous is lost; and the visual display is somewhat cluttered.

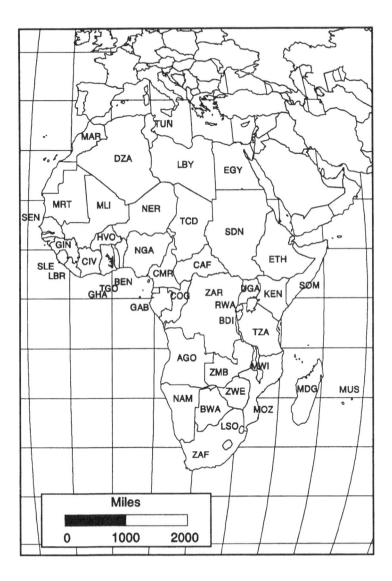

Figure 3.1. Africa, place names corresponding to those in Tables 3.1 and 3.2. Robinson projection is used as the base map.

TABLE 3.2 (source: World Bank)
COUNTRIES OF AFRICA, AGRICULTURE AND FOOD

	Val. Add. in Agr.		Cereal imports		Cereal aid		% Avg. food prod.	
	1974	1989	1974	1989	1974	1989	1987–1989 1979–1981=100%	
MOZ		704,000	62,000	400,000	34,000	424,000	83	Mozambique
ETH	931,000	2,254,000	118,000	690,000	54,000	573,000	89	Ethiopia
TZA	473,000	1,795,000	431,000	83,000	148,000	76,000	90	Tanzania
SOM	167,000	705,000	42,000	186,000	111,000	73,000	97	Somalia
MWI	119,000	498,000	17,000	86,000	0	217,000	85	Malawi
TCD	142,000	364,000	37,000	37,000	20,000	15,000	101	Chad
BDI	159,000	535,000	7,000	6,000	6,000	6,000	98	Burundi
SLE	108,000	409,000	72,000	145,000	10,000	38,000	89	Sierra Leone
MDG	243,000	717,000	114,000	103,000	7,000	76,000	93	Madagascar
NGA	5,080,000	8,874,000	389,000	240,000	7,000	0	96	Nigeria
UGA	929,000	2,986,000	36,000	16,000		17,000	87	Uganda
ZAR	805,000	2,846,000	343,000	323,000	1,000	55,000	94	Zaire
MLI	207,000	1,048,000	281,000	89,000	107,000	62,000	97	Mali
NER	420,000	744,000	155,000	105,000	73,000	83,000	86	Niger
HVO	121,000	871,000	99,000	120,000	28,000	49,000	115	Burkina Faso
RWA	135,000	799,000	3,000	10,000	19,000	2,000	77	Rwanda
KEN	484,000	2,208,000	15,000	119,000	2,000	112,000	101	Kenya
BEN	121,000	729,000	7,000	104,000	9,000	16,000	114	Benin
CAF	60,000	442,000	7,000	28,000	1,000	0	90	C. Afr. Rep.
GHA	1,030,000	2,570,000	177,000	244,000	33,000	46,000	109	Ghana
TGO	85,000	446,000	6,000	111,000	11,000	11,000	89	Togo
ZMB	191,000	617,000	93,000	123,000	5,000	66,000	97	Zambia
GIN		812,000	63,000	183,000	49,000	42,000	90	Guinea
LSO	23,000	83,000	48,000	140,000	14,000	34,000	80	Lesotho
MRT	58,000	339,000	115,000	207,000	48,000	70,000	88	Mauritania
LBR	91,000		42,000	158,000	3,000	28,000	95	Liberia
SDN	757,000		125,000	556,000	46,000	198,000	87	Sudan
AGO			149,000	248,000	0	79,000	84	Angola
EGY	1,942,000	5,858,000	3,877,000	8,543,000	610,000	1,427,000	109	Egypt
SEN	208,000	1,028,000	341,000	515,000	27,000	53,000	106	Senegal
ZWE	214,000	664,000	56,000	52,000	0	10,000	90	Zimbabwe
CIV	462,000	3,295,000	172,000	693,000	4,000	19,000	96	Cote d'Ivoire
MAR	789,000	3,679,000	891,000	1,329,000	75,000	238,000	120	Morocco
COG	49,000	311,000	34,000	82,000	2,000	2,000	98	Congo, PR
CMR	364,000	2,978,000	81,000	345,000	4,000	6,000	96	Cameroon
NAM		187,000		0			95	Namibia
TUN	245,000	1,235,000	307,000	1,655,000	59,000	284,000	96	Tunisia
BWA	28,000	75,000	21,000	77,000	5,000	33,000	68	Botswana
MUS	30,000	222,000	160,000	209,000	22,000	21,000	100	Mauritius
DZA	492,000	6,187,000	1,816,000	7,461,000	54,000	39,000	97	Algeria
ZAF	1,362,000	4,635,000	127,000	296,000			90	S. Africa
GAB	60,000	353,000	24,000	50,000			81	Gabon
LBY	93,000		612,000	1,515,000			109	Libya

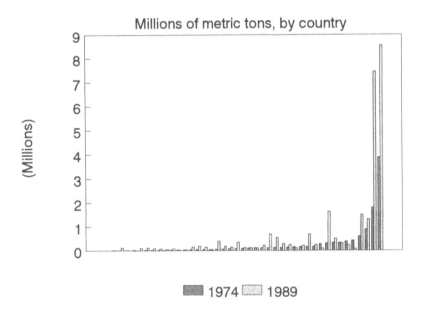

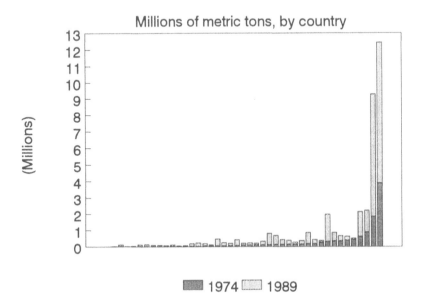

Figure 3.2 (top). Africa, cereal imports by country.
Figure 3.3 (bottom). Africa, cereal imports by country.

A "stacked" bar chart (Figure 3.3) produces a less cluttered display in which the discrete is still emphasized. However, even though this chart might have greater visual appeal than that in Figure 3.2, it does not present a clear picture of content. It is unknown what is the significance of adding together imports from 1974 and 1989--which is precisely what the "stacking" does. The chart in Figure 3.2 is therefore more appropriate than that in Figure 3.3, even though the latter is perhaps more attractive than the former.

Default Computer Curves

To focus on the continuous aspect of change in imports over time, graph the data as a curve in the x-y plane with country on the x-axis and millions of metric tons of cereal imports on the y-axis. To obtain an easy-to-read curve, arrange the data in one of the cereal imports columns in ascending order. There are two cereal imports columns, so there are two distinct ways to order the data; Table 3.3 displays data ordered in an ascending fashion for the 1974 cereal imports data. Table 3.4 displays data ordered on the 1989 column. In both tables, the first entry (Namibia) has a blank in the 1974 column. One might consider discarding this entry. We have retained it in the interest of completeness, insofar as the 1989 entry of reported data is 0. Decisions concerning issues of this sort need to be dealt with in some consistent manner.

When the 1974 curve is drawn as a line, and the 1989 curve is shaded, the visual display is relatively easy to read (Figures 3.4 and 3.5). The continuous aspect of change in imports is dominant; the discrete nature of the x-axis data is lost. With bar charts, it was easy to associate the positions of the bars with the countries in the table; with the continuous curves this is not as easy. Indeed, if one puts labels on the x-axis so that such an association is easier, then the labelling is likely to obscure the position of the curve which is often close to the x-axis (in fact, the x-axis and the bottom frame were removed from the graph in order to display more clearly the exact position of the curves). One could label the x-axis data; however, the price paid is too steep. Thus, a labelling of the first and last countries is chosen as a compromise. Graphs which involve both the discrete and the continuous will necessarily have problems of this sort associated with them; individuals will need to determine what tradeoffs are acceptable within the context in which they are working.

TABLE 3.3 (source: World Bank)
DATA ORDERED BY CEREAL IMPORTS, 1974

	Val. Add. in Agr.		Cereal imports		Cereal aid %		Avg. food prod.	
	1974	1989	1974	1989	1974	1989	1987–1989 1979–1981=100%	
NAM		187,000	0				95	Namibia
RWA	135,000	799,000	3,000	10,000	19,000	2,000	77	Rwanda
TGO	85,000	446,000	6,000	111,000	11,000	11,000	89	Togo
CAF	60,000	442,000	7,000	28,000	1,000	0	90	C. Afr. Rep.
BEN	121,000	729,000	7,000	104,000	9,000	16,000	114	Benin
BDI	159,000	535,000	7,000	6,000	6,000	6,000	98	Burundi
KEN	484,000	2,208,000	15,000	119,000	2,000	112,000	101	Kenya
MWI	119,000	498,000	17,000	86,000	0	217,000	85	Malawi
BWA	28,000	75,000	21,000	77,000	5,000	33,000	68	Botswana
GAB	60,000	353,000	24,000	50,000			81	Gabon
COG	49,000	311,000	34,000	82,000	2,000	2,000	98	Congo, PR
UGA	929,000	2,986,000	36,000	16,000		17,000	87	Uganda
TCD	142,000	364,000	37,000	37,000	20,000	15,000	101	Chad
LBR	91,000		42,000	158,000	3,000	28,000	95	Liberia
SOM	167,000	705,000	42,000	186,000	111,000	73,000	97	Somalia
LSO	23,000	83,000	48,000	140,000	14,000	34,000	80	Lesotho
ZWE	214,000	664,000	56,000	52,000	0	10,000	90	Zimbabwe
MOZ		704,000	62,000	400,000	34,000	424,000	83	Mozambique
GIN		812,000	63,000	183,000	49,000	42,000	90	Guinea
SLE	108,000	409,000	72,000	145,000	10,000	38,000	89	Sierra Leone
CMR	364,000	2,978,000	81,000	345,000	4,000	6,000	96	Cameroon
ZMB	191,000	617,000	93,000	123,000	5,000	66,000	97	Zambia
HVO	121,000	871,000	99,000	120,000	28,000	49,000	115	Burkina Faso
MDG	243,000	717,000	114,000	103,000	7,000	76,000	93	Madagascar
MRT	58,000	339,000	115,000	207,000	48,000	70,000	88	Mauritania
ETH	931,000	2,254,000	118,000	690,000	54,000	573,000	89	Ethopia
SDN	757,000		125,000	556,000	46,000	198,000	87	Sudan
ZAF	1,362,000	4,635,000	127,000	296,000			90	S. Africa
AGO			149,000	248,000	0	79,000	84	Angola
NER	420,000	744,000	155,000	105,000	73,000	83,000	86	Niger
MUS	30,000	222,000	160,000	209,000	22,000	21,000	100	Mauritius
CIV	462,000	3,295,000	172,000	693,000	4,000	19,000	96	Cote d'Ivoire
GHA	1,030,000	2,570,000	177,000	244,000	33,000	46,000	109	Ghana
MLI	207,000	1,048,000	281,000	89,000	107,000	62,000	97	Mali
TUN	245,000	1,235,000	307,000	1,655,000	59,000	284,000	96	Tunisia
SEN	208,000	1,028,000	341,000	515,000	27,000	53,000	106	Senegal
ZAR	805,000	2,846,000	343,000	323,000	1,000	55,000	94	Zaire
NGA	5,080,000	8,874,000	389,000	240,000	7,000	0	96	Nigeria
TZA	473,000	1,795,000	431,000	83,000	148,000	76,000	90	Tanzania
LBY	93,000		612,000	1,515,000			109	Libya
MAR	789,000	3,679,000	891,000	1,329,000	75,000	238,000	120	Morocco
DZA	492,000	6,187,000	1,816,000	7,461,000	54,000	39,000	97	Algeria
EGY	1,942,000	5,858,000	3,877,000	8,543,000	610,000	1,427,000	109	Egypt

Different graphs emerge in Figures 3.4 and 3.5 even though the data are the same. When data are sorted on 1974 cereal imports (Table 3.3), lines from 1974 may not match the 1989 sorting (Table 3.4). Both views are "historical" in nature; they compare the 1974 view to the 1989 view. When the data are ordered on the 1974 column, the curve generated by that data is smoother than the 1979 curve (Figure 3.4). When the data are ordered on the 1989 column, the curve generated by that data is smoother than the 1974 curve (Figure 3.5). Smooth-looking curves tend to dominate in visual displays. If the 1974 data is to dominate (to visualize cumulative, evolutionary change) then the data should be ordered on the earlier of the two data sets.

TABLE 3.4 (source: World Bank)
DATA ORDERED BY CEREAL IMPORTS, 1989

	Val. Add. in Agr.		Cereal imports		Cereal aid		% Avg. food prod.	
	1974	1989	1974	1989	1974	1989	1987–1989	
							1979–1981=100%	
NAM		187,000		0			95	Namibia
BDI	159,000	535,000	7,000	6,000	6,000	6,000	98	Burundi
RWA	135,000	799,000	3,000	10,000	19,000	2,000	77	Rwanda
UGA	929,000	2,986,000	36,000	16,000		17,000	87	Uganda
CAF	60,000	442,000	7,000	28,000	1,000	0	90	C. Afr. Rep.
TCD	142,000	364,000	37,000	37,000	20,000	15,000	101	Chad
GAB	60,000	353,000	24,000	50,000			81	Gabon
ZWE	214,000	664,000	56,000	52,000	0	10,000	90	Zimbabwe
BWA	28,000	75,000	21,000	77,000	5,000	33,000	68	Botswana
COG	49,000	311,000	34,000	82,000	2,000	2,000	98	Congo, PR
TZA	473,000	1,795,000	431,000	83,000	148,000	76,000	90	Tanzania
MWI	119,000	498,000	17,000	86,000	0	217,000	85	Malawi
MLI	207,000	1,048,000	281,000	89,000	107,000	62,000	97	Mali
MDG	243,000	717,000	114,000	103,000	7,000	76,000	93	Madagascar
BEN	121,000	729,000	7,000	104,000	9,000	16,000	114	Benin
NER	420,000	744,000	155,000	105,000	73,000	83,000	86	Niger
TGO	85,000	446,000	6,000	111,000	11,000	11,000	89	Togo
KEN	484,000	2,208,000	15,000	119,000	2,000	112,000	101	Kenya
HVO	121,000	871,000	99,000	120,000	28,000	49,000	115	Burkina Faso
ZMB	191,000	617,000	93,000	123,000	5,000	66,000	97	Zambia
LSO	23,000	83,000	48,000	140,000	14,000	34,000	80	Lesotho
SLE	108,000	409,000	72,000	145,000	10,000	38,000	89	Sierra Leone
LBR	91,000		42,000	158,000	3,000	28,000	95	Liberia
GIN		812,000	63,000	183,000	49,000	42,000	90	Guinea
SOM	167,000	705,000	42,000	186,000	111,000	73,000	97	Somalia
MRT	58,000	339,000	115,000	207,000	48,000	70,000	88	Mauritania
MUS	30,000	222,000	160,000	209,000	22,000	21,000	100	Mauritius
NGA	5,080,000	8,874,000	389,000	240,000	7,000	0	96	Nigeria
GHA	1,030,000	2,570,000	177,000	244,000	33,000	46,000	109	Ghana
AGO			149,000	248,000	0	79,000	84	Angola
ZAF	1,362,000	4,635,000	127,000	296,000			90	S. Africa
ZAR	805,000	2,846,000	343,000	323,000	1,000	55,000	94	Zaire
CMR	364,000	2,978,000	81,000	345,000	4,000	6,000	96	Cameroon
MOZ		704,000	62,000	400,000	34,000	424,000	83	Mozambique
SEN	208,000	1,028,000	341,000	515,000	27,000	53,000	106	Senegal
SDN	757,000		125,000	556,000	46,000	198,000	87	Sudan
ETH	931,000	2,254,000	118,000	690,000	54,000	573,000	89	Ethopia
CIV	462,000	3,295,000	172,000	693,000	4,000	19,000	96	Cote d'Ivoire
MAR	789,000	3,679,000	891,000	1,329,000	75,000	238,000	120	Morocco
LBY	93,000		612,000	1,515,000			109	Libya
TUN	245,000	1,235,000	307,000	1,655,000	59,000	284,000	96	Tunisia
DZA	492,000	6,187,000	1,816,000	7,461,000	54,000	39,000	97	Algeria
EGY	1,942,000	5,858,000	3,877,000	8,543,000	610,000	1,427,000	109	Egypt

If the current view is to dominate, then order the data on the most current data. The needs of the particular real-world problem should dictate what viewpoint dominates. Each of these views used only the simplest default settings available. A number of issues should be considered even in using simple default settings. Generally, it is also not much more difficult to introduce various other analytic methods for examining observed patterns in data. Let the problem at hand guide the choice of tool, so that tool and problem fit.

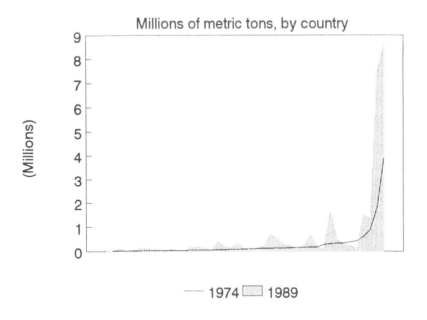

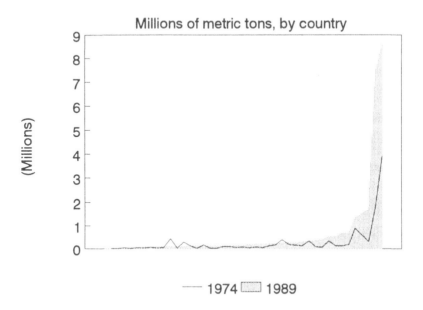

Figure 3.4 (top). Africa, cereal imports by country, 1974 (Table 3.3).
Figure 3.5 (bottom). Africa, cereal imports by country, 1989 (Table 3.4).

Linear and non-linear fitting: some unbounded curves

All of the default spreadsheet settings examined above involved using bounded, linear curves to fit data. These curves were not smooth, but they did suggest, quite well, the general patterns in the data. Thus, there seems no point to refining the graphs by smoothing the curves. Smoothing of bounded curves is of particular use when interpolation between observed values is important. In agricultural data of the sort in Table 3.1, interpolation does not seem to make sense.

What is perhaps of greater interest, is to consider how, given past and present agricultural trends, one might project them into the future. Projection of this sort is of course impossible to do "correctly;" there are an infinite number of curves that fit any finite distribution of points. Nonetheless, projections are made all the time, particularly to suggest policy directions. A guess about the future based on some knowledge of previous trends carries more weight than does blind speculation; however, the projections must also make sense--any two data sets can be compared, but often not in a sensible manner.

The graphs in Figures 3.4 and 3.5 suggest that policy makers might attempt to fit an exponential curve to cereal imports for Africa in order to forecast future needs. The dark line tracery in Figure 3.4 (ordered on 1974 data) suggests an exponential curve, as does the shaded curve in Figure 3.5 (ordered on 1989 data). Indeed, exponential curves can be "fit," and when they are, the rate of increase in the curve is steeper than the corresponding rate of increase in the exponential fit to the given data using least squares to fit the exponential. This fact could lead to rash predictions about the state of cereal imports in Africa. What the careful user of curves must note, however, is that the inputs on the x-axis in Figures 3.4 and 3.5 are countries; it is meaningless to try to interpolate between them, and it is also meaningless to try to extrapolate; there are 42 countries in this data set. To extrapolate the apparently exponential curve in Figures 3.4 and 3.5 would require extrapolating the number of countries, as well. Extrapolation through space is generally not well-defined.

Thus, with data of this sort, sequences of data gathered at evenly-spaced time intervals (sometimes called "time series"--a mathematical misnomer) are useful in extrapolating through time. The World Bank data has entries only for 1974 and 1989; two entries are not sufficient for considering numerical differences; to do so, tables in a different data bank were considered. Using multiple data bases is an important strategy; a data base that is excellent for one purpose may not be useful for another. In addition, the use of multiple sets permits the user to check easily the consistency of the entries.

TABLE 3.5 (source: World Bank)
CEREAL IMPORTS IN THOUSAND TONS, AFRICA
Arranged in increasing numerical order on the 1989 column

	1970	1971	1972	1973	1974	1975	1976	1977	1978	1979	1980	1981	1982	1983	1984	1985	1986	1987	1988	1989
BDI	12.4	10.5	13.4	6.1	7.3	8.6	11.6	10	14.5	15.8	18.2	18.4	20.4	20.2	13.6	19.5	13.1	12.7	14.9	5.9
RWA	12.8	6.3	6.5	7.7	3	17.3	9.5	12.2	14.6	6.5	16.4	15.3	14.1	24.1	25	32.8	27.1	12.5	8.8	9.8
UGA	40.5	27.9	61	28.9	36.5	6.2	10.4	8.6	0	10.1	52.2	40.4	68.1	23.1	30.6	20.2	16.7	20.1	22	16.1
CAF	10.4	13	9.4	11.8	6.7	10.4	9.9	8.6	3.5	10.9	11.8	18.9	15.5	15	12.6	19.8	39.9	34.6	40.1	28.4
TCD	10.1	9.4	9.5	19.9	36.7	10.4	35.5	19.9	18.2	21.7	15.6	30.1	57.2	31.2	84.6	86.1	36.3	57.1	55.9	37.3
GAB	12.8	11.3	11.5	12.3	24.2	62.9	42.3	40.5	26.2	28.9	26.7	34.7	43.5	48.3	55.2	63.6	52.9	55.2	58.5	50.2
ZWE	68	98	50	39.1	56.2	23.9	27.5	37.8	39.3	9.8	155.5	28.5	11.3	74.8	370.9	152.9	54.3	47.2	93	52
BWA	70.2	38.7	41.7	35.9	20.6	57.1	32.2	54.6	39.3	75.7	68.3	49.1	61.1	122.9	174.3	154.6	112.6	94.3	106.7	76.7
COG	24.9	36.2	37.1	30.8	34.3	35.7	32.2	54.6	50.8	63.9	88.2	50.6	80.6	79.9	107.9	99.7	93.5	78.2	113.2	82
TZA	67.8	57.4	183.6	42.8	430.7	461	97.4	134.8	123.4	58.9	398.6	266.1	324.5	232.4	269.6	402	222.4	179	120.2	82.7
MWI	113.9	28.5	28.1	26.9	17	41.3	43.2	27	13.7	14.8	36.3	74.2	26.5	20.1	20.9	29.4	15.8	60.6	42	86.2
MLI	31.2	63.4	64.3	192.5	281.3	119.8	30.4	24.1	22	239.1	86.8	74.2	177	171.5	253.5	313.2	192.7	75.2	112.3	89.3
MDG	54.8	99.8	98.3	104.1	114.4	72.2	78.1	135.9	185.1	239.1	110.1	279.4	391.4	264.6	162.5	195.1	211.9	177.7	86.4	103.3
BEN	14.7	24.1	41.6	17.9	7.5	12.3	35.5	64.8	22	44.7	61.3	88.6	148.6	114.1	89.5	66.6	72.1	76.8	130.7	104.4
NER	12.9	8.1	19.4	60.3	155.5	17.1	62.2	28.3	69.9	44.7	63	61	56.1	128.2	86.7	258.8	58.6	64.6	151.4	104.5
TGO	20.1	15.9	20.6	20.3	6	2	12	51.9	26.5	38.2	40.8	149.1	89.4	80.9	91	57	60.5	86.1	136	111.4
KEN	26.5	62.8	71.8	81.1	15.3	86.1	11.6	34.3	99.7	38.2	386.7	250.4	205.6	209.8	556.4	279.3	178.7	257	85.6	119
HVO	29.9	26.8	25.2	39.9	98.8	25.9	28.6	54.4	70	80.7	77.1	48.1	89.4	82	148	200.9	123.1	93.5	129.8	119.7
ZMB	143.4	353.7	185.8	86.5	93.1	164	104.7	105.5	100.2	225.2	498.4	149.1	127.9	59.3	239.9	200.6	164.7	135.1	118.6	123
LSO	39.6	41.8	40	57.8	48.5	56.2	85.6	82.3	95.2	95.3	107.4	100.3	116.6	101.7	129.6	119	131.7	78.7	146	139.8
SLE	87.9	59.2	52.1	80.2	71.8	25.8	32.5	45.7	49	127.3	83.5	77.7	94	59.3	53	118.9	123.6	147.9	119	145.1
LBR	58.4	64.4	30.9	56.4	41.7	41.5	47	67.1	77	97	99.2	110.9	111.7	101.7	161.7	103.5	117.7	100	101.9	157.5
GIN	47.4	28.9	68.4	75.9	63.2	67.3	57.8	59	89.8	109.8	171.4	130.4	116.6	249	263.8	140.2	150.8	172	178.1	183.3
SOM	56.1	126.4	91.8	50.1	42.2	169.4	148.1	185.9	66.3	136.5	220.6	430	393.8	300.4	264.8	179.8	257.7	301.6	211.9	185.6
MRT	64.9	78.1	144.1	146.2	114.9	107.7	133.1	150.8	161.9	132	165.9	141.3	152	167.5	187.6	290.9	168.5	189	213.4	206.9
MUS	126.4	118	380.6	124.6	160.3	149.4	136.7	140.5	158.4	151.1	180.6	175.2	178.6	1475.	1529	156.9	162	188.8	176.4	209.3
NGA	327.9	467.3	112.8	451.2	389.3	447.5	823.6	1320.	2007.	1652.	1827.	2215.	2158.	1475.	1529	1956.	1368.	594.6	339.3	240
GHA	143.7	84.9	94.2	110.5	177	163.3	104.6	176	324.7	164.3	247.3	162.9	171.1	230	192.7	136.9	144.7	223.4	256.7	243.9
AGO	99	88.9	109.5	140.9	127.2	96	94.1	137.7	117.5	171.2	159.4	288.4	290.7	287.1	325.9	284.4	158.6	279.8	292	247.8
ZAF	421.1	161.4	243.3	330.5	343	302	122.2	200.1	240.1	334.5	350.1	480.4	301.9	1516.	3259.	763.4	731	338.4	255.5	295.5
ZAR	205.2	251.6	96.7	91.2	81.1	68.8	385.4	348.6	123.6	163.6	140.3	551.6	333.3	3226.	291	319.7	408.3	475.9	415	322.6
CMR	91.3	112.9	88.3	126.5	62.3	81.1	74.5	116.2	123.6	353.4	368.5	107.2	117.1	202	122.1	141.1	190.9	289.5	349.9	345.4
MOZ	111.2	111.5	287.8	457.6	340.9	191.1	194.7	165.1	221	353.4	368.5	319.9	307.2	274.4	429.8	598.3	244.1	373.7	526.8	400
SEN	239.4	365.1	222	203.1	125.3	220.1	429.5	415.3	453.9	502.8	451.9	484.7	494.2	544.4	661.5	496	511.8	421.9	458.1	515
SDN	240.1	194.3	9.2	19.2	117.6	124.3	122.5	136.9	172.6	222.6	236.2	295	437.4	451.8	518.4	1147.	650.9	739.3	702	556.1
ETH	77	47.7	154.1	289.6	172	67	92	165.1	217.8	250.1	397.4	209.6	278	344.6	2522.	716.6	963.8	579.1	1126.	689.5
CIV	183.9	130.7	478.7	1059.	890.9	81.7	122.3	282.1	307.7	360.8	469.3	572.8	541.2	597.1	535.5	553.9	579.6	760.4	495.6	692.7
MAR	419.1	772.9	301.6	798.9	612	1508.	1223.	1419.	1689.	1654	1820.	2723.	1975.	2014.	2769.	2176.	1609.	2235.	1643.	1329.
LBY	377.4	440.5	340.8	306.9	337.3	600.7	474.9	671.2	653.9	532.4	907.7	683.7	975.6	8678.	1011.	1113.	1784.	1425.	1435.	1514.
TUN	486.5	350.7	293.3	172	890.9	337.3	414.9	723.3	792.5	897.5	817	947.8	1026	1133.	1073.	731.6	1312.	1167.	2116.	1654.
DZA	359.3	766.5	1289.	860.2	1815.	1668.	1857.	2110.	3136.	2974.	3413.	2719.	4094.	3766.	4115.	5266.	4610.	3865.	5368	7460.
EGY	1305.	2247.	1773.	1872.	3877.	4214	4347.	4935.	5852.	5400.	6027.	7199	6799.	8114.	8616.	8903.	8407.	9348.	8500.	8542.

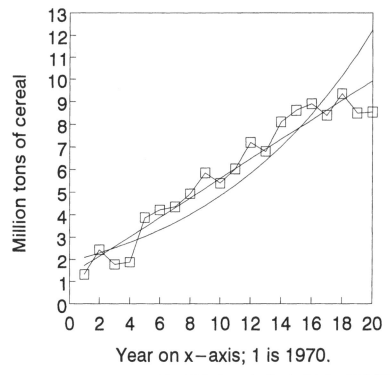

Figure 3.6. Egypt, cereal imports 1970-1989 (in thousands of tons). Data from Table 3.5. Graph shows default fit of connecting the data points (squares) and an exponential and linear fit to the data.

Data sets of the sort displayed in Table 3.5, derived from the World Resources Institute Data Base (WRD), are useful in looking for trends over time at any scale for which there is data. This table, with data recorded on an annual basis, may be used effectively to compare import trends between countries over a fifteen year span, or simply to consider the general trend of a single country's cereal import pattern. Because agriculture is generally based on a yearly cycle (although not always) it would not make sense to interpolate values between the annual values; the data is already gathered with as small a spacing between data points as is reasonable (with respect to agriculture). Because the set is gathered over time, it does make sense to extrapolate from this set to understand how the trend of the actual data might be projected into the future.

A general view of the data in Table 3.5 suggests an overall increase in cereal imports, for most countries in the list, over the 20 year time span for which there is data. The straightforward approach of considering the import patterns of one country will illustrate the importance of understanding curve fitting if one is to make projections, and allocate funding or resources, on the basis of the curves from which the projections are derived.

Thus, consider the case of Egypt, which imported about 13 million tons of cereal in 1970 and about 85 million tons in 1989. In a spreadsheet, the default graph of the data, as an xy graph with year on the x-axis and thousands of tons of cereal imports on the y-axis, is simply to connect the 20 data points (one for each year) with 19 line segments (Figure 3.6). While this strategy offers a useful graphic view of the data, it does not offer any consistent way to make a projection based on the data. To do so, fit some mathematical function, suggested by the general pattern of the data, to the existing data and extrapolate that curve beyond the given data set.

To fit a straight line to the data, proceed as follows (refer to Chapter 1 to understand why this procedure works).

STRAIGHT LINE FIT TO THE DATA
(Refer to Table 3.6 and Figure 3.6)

1. Enter the years in a spreadsheet column (B) to be used as the x-axis input.
2. Enter the data that varies over time in another column (spreadsheet column C--cereal imports in this case), to be used as y values
3. Choose the regression feature from the software, with the x values as in step 1 and the y values as in step 2.
4. Choose the output range as a blank area in the spreadsheet. Then proceed with the calculation as directed by the software; the output from the regression will appear in a form similar to the one in Table 3.7, top half (produced in Lotus 1-2-3, rel. 2.3).
5. The equation below the output range, set in bold 12 point type, must generally be derived by the user from the regression output. The slope-intercept form for the equation of a straight line ($y=mx+b$) is used. The "X Coefficient" from the regression output is used as "m". The "Constant" from the regression output is used as b. Thus, the equation can be read directly from the regression output.

TABLE 3.6

A	B	C	D	E	F
	x	y	LN y	proj lin	proj exp
1	1970	1305.7	7.174494574	1723.17	2094.688219
2	1971	2447.8	7.802944940	2154.931	2298.674657
3	1972	1773.6	7.480766658	2586.692	2522.525850
4	1973	1872.2	7.534869488	3018.453	2768.176280
5	1974	3877.3	8.262894314	3450.214	3037.748819
6	1975	4214	8.346167594	3881.975	3333.573064
7	1976	4347.3	8.377310241	4313.736	3658.205479
8	1977	4935.4	8.504189002	4745.497	4014.451482
9	1978	5852.6	8.674641285	5177.258	4405.389690
10	1979	5400.4	8.594228303	5609.019	4834.398525
11	1980	6027.8	8.704137380	6040.78	5305.185407
12	1981	7199	8.881697406	6472.541	5821.818797
13	1982	6799.8	8.824648479	6904.302	6388.763352
14	1983	8114.3	9.001383216	7336.063	7010.918510
15	1984	8616.2	9.061399431	7767.824	7693.660830
16	1985	8903.5	9.094199736	8199.585	8442.890455
17	1986	8407.4	9.036867549	8631.346	9265.082100
18	1987	9348.5	9.142971181	9063.107	10167.34100
19	1988	8500.3	9.047856736	9494.868	11157.46433
20	1989	8542.7	9.052832396	9926.629	12244.00857
21	1990			10358.39	13436.36343
22	1991			10790.151	14744.83306
23	1992			11221.912	16180.72503
24	1993			11653.673	17756.44806
25	1994			12085.434	19485.61928
26	1995			12517.195	21383.18191
27	1996			12948.956	23465.53434
28	1997			13380.717	25750.67192
29	1998			13812.478	28258.34242
30	1999			14244.239	31010.21669
31	2000			14676	34030.07597
32	2001			15107.761	37344.01736
33	2002			15539.522	40980.67938
34	2003			15971.283	44971.48944
35	2004			16403.044	49350.93544
36	2005			16834.805	54156.86381
37	2006			17266.566	59430.80656
38	2007			17698.327	65218.34021
39	2008			18130.088	71569.47963
40	2009			18561.849	78539.11029
41	2010			18993.61	86187.46254
42	2011			19425.371	94580.63215
43	2012			19857.132	103791.1514
44	2013			20288.893	113898.6161
45	2014			20720.654	124990.3733
46	2015			21152.415	137162.2759
47	2016			21584.176	150519.5116
48	2017			22015.937	165177.5111
49	2018			22447.698	181262.9466
50	2019			22879.459	198914.8256
51	2020			23311.22	218285.6925
52	2021			23742.981	239542.9472
53	2022			24174.742	262870.2911
54	2023			24606.503	288469.3153
55	2024			25038.264	316561.2421
56	2025			25470.025	347388.8372

6. In a separate column of the spreadsheet, labelled "proj lin" in Table 3.6 (Column E), enter the equation derived from the regression: in cell E2, in the first row under the label "proj lin" (second row of the spreadsheet), enter the formula 431.7610*B2-848846. The value 1723.17 should appear at the top of the "proj lin" column. Then, copy the cell content from E2 to the 55 cells below it; this should produce the entire numerical range shown in the "proj lin" column.

7. Graph the results. On the default graph, enter the first 20 years of the projected linear data in an additional range of the spreadsheet. To produce much of the graph in Figure 3.6, the first 20 years of column A were entered as x-axis labels in the x-range; then 20 years of y values (column C) were entered in the "A" range of Lotus 1-2-3; and then, 20 years of projected linear data (column E) were entered in the "B" range of Lotus 1-2-3 to display the linear fit to the data. In the latter case, only lines, and not symbols were displayed (again for clarity). See Figure 3.6.

The error terms in the straight line fit suggest that one might not place much confidence in this fit. Any of an infinite number of curves may be used to fit the data; the art comes in making reasonable selections. Current computers are excellent at making the fits; thus, the place to question people about curves fit to data is in their choice of curve used to fit data--particularly when projections are made from that fit. Thus, a better fit to the Egyptian data is sought.

BLACK BOX SUMMARY
see Introduction for theoretical explanation

LEAST SQUARES REGRESSION LINE

$$y=mx+b$$

where
m is the slope of the line, or the "x-coefficient"
b is the y-intercept, or the "constant."

EXPONENTIAL CURVE FIT TO THE DATA
(Refer to Table 3.6 and Figure 3.6)

1. Enter the years in a spreadsheet column (B) to be used as the x-axis input.

2. Enter the data that varies over time in another column (spreadsheet column C--cereal imports in this case), to be used as y values.

3. Take the natural log of the y values (other bases for the logarithms work, too).

4. Choose the regression feature from the software, with the x values as in step 1 and the y values as in step 3--the regression will be run on the x values and the LN y values (to obtain the exponential form).

5. Choose the output range as a blank area in the spreadsheet. Then proceed with the calculation as directed by the software; the output from the regression will appear in a form similar to the one in Table 3.7, bottom half (produced in Lotus 1-2-3, release 2.3).

6. The equations below the output range, set in bold 12 point type, must generally be derived by the user from the regression output. The slope-intercept form for the equation of a straight line (LN y=mx+b) is used. The "X Coefficient" from the regression output is used as "m". The "Constant" from the regression output is used as b. Thus, the first equation can be read directly from the regression output.

7. The second equation, that is an exponential, is derived from the first equation by raising both sides to e, the base of the natural logarithms. This is the equation that will be used to enter projected exponential values.

8. In a separate column of the spreadsheet, labelled "proj exp" in Table 3.6 (Column F), enter the equation derived from the regression: in cell F2, in the first row under the label "proj lin" (second row of the spreadsheet), enter the formula (in Lotus 1-2-3 format--adjust for other software) @EXP(-175.421)*@EXP(0.092928*(B2)). The value 2093.688219 should appear at the top of the "proj lin" column. Then, copy the cell content from F2 to the 55 cells below it; this should produce the entire numerical range shown in the "proj lin" column.

TABLE 3.7
Regression Output:

Constant	−848846.71
Std Err of Y Est	685.575621
R Squared	0.93611431
No. of Observations	20
Degrees of Freedom	18

X Coefficient(s)	431.761052
Std Err of Coef.	26.5854822

y=431.7610x−848846

Regression Output:

Constant	−175.42103
Std Err of Y Est	0.24794458
R Squared	0.83843853
No. of Observations	20
Degrees of Freedom	18

X Coefficient(s)	0.09292804
Std Err of Coef.	0.00961487

LN y = 0.092928x −175.421
y=e^(−175.421)*e^(0.092928x)

9. Graph the results. On the default graph, enter the first 20 years of the projected exponential data in an additional range of the spreadsheet. To produce much of the graph in Figure 3.6, the first 20 years of column A were entered as x-axis labels in the x-range; then 20 years of y values (column C) were entered in the "A" range of Lotus 1-2-3; and then, 20 years of projected linear data (column E) were entered in the "B" range of Lotus 1-2-3 to display the linear fit to the data; now, enter 20 years of projected exponential data (column F) in the C range of Lotus 1-2-3 to display the exponential fit of the data. In the latter two cases, only lines, and not symbols were displayed (again for clarity). See Figure 3.6.

The error terms suggest that the exponential curve offers a better fit than did a straight line to the given data; no doubt even better fits can be found using different curves. Even with just these two curves, consider how they might be used to make projections: each appears fairly close to the data, and not far apart from one another. How will they appear in the year 2000, or in the year 2025? Figure 3.7 shows the anwers to these questions. It may not make much difference, NOW, which curve one chooses to fit to the data; but, when the curve is extrapolated, and when funding allocations for cereal imports are planned for the future, it can make a big difference which of these seemingly similar curves is used as a tool.

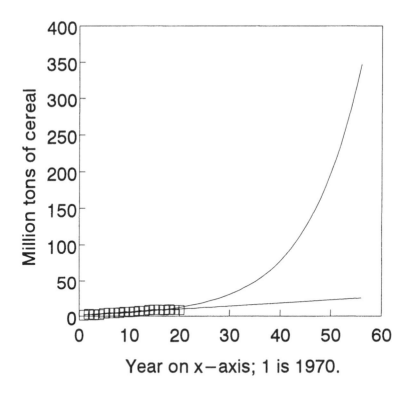

Figure 3.7. Egypt, cereal imports 1970-1989 (in thousands of tons) shown as lines connected to the data points (squares). The exponential and linear curves fit to the actual data are extrapolated to the year 2025 (Table 3.6).

EXTRAPOLATION

To extrapolate the linear and exponential curves fit to the actual data, simply graph the entire lengths of columns A, E, and F, on the same graph--thus, Figure 3.6 becomes Figure 3.7.

BLACK BOX SUMMARY
see Introduction for theoretical explanation

EXPONENTIAL CURVE

$$y = Ce^{ax} + b$$

where

$$a < 0$$

and

$y=b$ is the lower bound of the exponential;
C is a constant.

Of course, the world's cereal supply is not infinite, so that cereal imports clearly cannot follow an exponential model indefinitely. Thus, one expects to see some upper bound as a limit to the amount of cereal that could be imported; when this view is taken, the curve that apparently began as an exponential, tapers off as a logistic curve, approaching the upper bound from below as an asymptote. Data in other chapters will illustrate the mechanics of fitting a logistic curve to real world data.

What we can see, just from examining the straight line fit and the exponential fit to the Egyptian data is that now is the time to guide the direction of the future. If the current pattern continues exponentially, trouble will follow. The intersection point (farthest to the right) of the straight line and the exponential pinpoints where the takeoff to disaster begins; it is here, that alternative futures can be planned for, to follow any of a variety of trajectories.

REFERENCES

1. Ness, G., Drake, W., and Brechin, S. Population-Environment Dynamics: Ideas and Observations, The University of Michigan Press, Ann Arbor, Michigan, 1993.

CHAPTER 4

BIODIVERSITY DATA ANALYSIS

ANALYTICAL TECHNIQUES/TOOLS USED

Dot Maps
Equal-Area Map Projections
Geographic Information Systems
Map Overlays
Feigenbaum's Graphical Analysis

DATA TYPE: BASELINE, ONLY, AVAILABLE

Overview of Data

The diversity of biological species, or its lack, is one way to measure the health of various elements of the Earth's environment (Kates and Burton 1986). Heterogeneity is security against extinction. Biodiversity is a tool for examining global change that is only presently becoming a widely-used indicator. Data on species distribution, at any geographical scale, are difficult to collect; a census of humans is a major effort-- imagine the difficulty in taking a census of frog species or plant taxa! Because of the extreme difficulty in accumulating information, data sets concerning biodiversity are unusually hard to find, and because the demand for them is only recent, longitudinal data, across a sequence of years, is even harder to find.

The World Resources Institute data base contains data on numbers of species of mammals, birds, freshwater fish, amphibians, and taxa of plants. It has separate tables for known species, threatened species, and threatened species per unit of land area. Data is available for a single year only for each of these indicators of global biodiversity. Thus, these data sets serve as good models of how to deal with baseline data sets--ones that will serve as the base, or starting point, for data that is to be accumulated over a sequence of years to come.

With any data set (presented in electronic or paper format), it is important first to examine the set for interesting or unusual patterns in the display. These patterns often influence decisions in choosing data and analytical tools.

PATTERNS IN DATA--WHAT TO LOOK FOR

1. What is the general organizational scheme of the entire set? Is it arranged alphabetically, numerically, or in some other fashion?

2. Are the real-world entries in the Table (nations, states, counties) expressed as comparable units? For example, county data and national data are generally not comparable.

3. Are the numerical entries in the Table expressed in comparable units? For example, data in one column might measure percentages while data in another column might measure thousands of dollars--these columns would not be comparable.

4. Are there gaps in the data? If so, what is their significance to the questions you wish to have the data answer?

Using these four items as a guide, consider the data set displayed in Table 4.1. This table displays a set of data by country showing the number of known amphibian species in 1989 (in columns 2 and 3). The data set is "clean"--there are no gaps and the units are numerically comparable. Note in particular that the number of known amphibian species per unit of land, in the column farthest to the right, is expressed per 10,000 kilometers squared. It could easily be expressed in terms of any number of kilometers squared; had it been written per square kilometer instead, there would be substantial numerical clutter from all the zeroes introduced in representing ten-thousandths of a unit. In its present form, this column is easy to read and numerical rankings are easy to grasp. In addition, other entries in the electronic data base are expressed per ten-thousand kilometers squared, so that this column will then become compatible with other existing ones, too.

TABLE 4.1: KNOWN AMPHIBIAN SPECIES
AND OTHER DATA
(Sources: World Resources Institute Data Base;
Goode's World Atlas)

Latitude	Country	Known Amphib. Species	Area km. sq.	Known amphib. spec. per 10,000 km. sq.
50.72	Canada	41	9,976,000	0.0410986367
59.58	Finland	5	338,000	0.1479289941
63.8	Norway	5	324,000	0.1543209877
−39.23	New Zealand	5	269,000	0.1858736059
−37.17	Australia	150	7,687,000	0.1951346429
8.95	Nigeria	19	924,000	0.2056277056
38.75	Turkey	18	779,000	0.2310654685
38	United States	222	9,373,000	0.2368505281
60.17	Sweden	13	450,000	0.2888888889
−35.5	Argentina	124	2,767,000	0.4481387785
40.25	Spain	24	505,000	0.4752475248
−35	Chile	38	757,000	0.5019815059
46.65	France	29	552,000	0.5253623188
23	India	181	3,288,000	0.550486618
−9	Brazil	487	8,512,000	0.5721334586
−22.17	Botswana	38	582,000	0.6529209622
32	Morocco	32	447,000	0.7158836689
−28	South Africa	95	1,221,000	0.7780507781
−17	Bolivia	96	1,099,000	0.8735213831
43.97	Italy	28	301,000	0.9302325581
−14.38	Zambia	83	753,000	1.1022576361
23.75	Mexico	284	1,958,000	1.4504596527
1	Kenya	88	580,000	1.5172413793
−24	Paraguay	69	407,000	1.6953316953
−10	Peru	235	1,285,000	1.8287937743
14.42	Philippines	60	300,000	2
8	Venezuela	183	912,000	2.0065789474
18	Vietnam	80	330,000	2.4242424242
36.5	Japan	95	378,000	2.5132275132
−17.83	Zimbabwe	120	391,000	3.0690537084
3.5	Colombia	375	1,139,000	3.2923617208
0	Ecuador	350	284,000	12.323943662

Geographical tools--maps

One way to slice through data sets is to use time; another way is to use space. Time is not available as a way to sort data in this case, so in baseline cases, there is generally a total reliance on space.

To get a good view of the global distribution of species, it is useful to visualize them on a map. On Figure 4.1, each dot represents 3 known amphibian species. Thus, it is easy to estimate the number of known amphibian species contained in any region--if a region contains 20 dots, it contains no more than 60 known species of amphibians--it might contain substantially fewer if the region crosses national political boundaries and two different dots represent the same sets of known amphibian species. The problem of overlap needs to be considered.

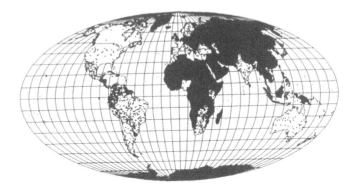

Figure 4.1. Distribution of known amphibian species for a set of countries--1 dot represents three species. Mollweide equal-area projection used as base map. Landmasses for which there is no data in this set are shaded.

Analysis using dot maps, in which such counts can be made independent of the political boundaries, offers a way to obtain rough information at sub-national levels even when that data is not directly available in the original data set. One must be careful though; the dots are always randomly placed within the smallest unit for which data was gathered (may not be the smallest unit displayed). Thus, dot maps are not good for estimating distributions in very small regions (where "very small" is related to the size of the smallest unit for which data was actually collected).

The map projection chosen in this case is a Mollweide projection--an equal area projection. This sort of projection is useful for comparing areas; a unit square in Greenland represents the same amount of territory as the same size unit square in Brazil--in contrast to the widely used Mercator map, which really should be used only for navigational (and similar) purposes. When distributions across areas are considered, use an equal area projection to map data.

Many software mapping packages (Geographic Information Systems--GISs) can import data and analyze and map the results; when using such packages, be sure to choose a projection suited to the display. Often the default is simply based on an evenly-spaced grid of latitude and longitude lines which produces a badly distorted map with greatly exaggerated landmasses toward the poles (Figure 4.2). Dot maps are easy to make using software of this sort; what is important is to consider elements of the base map, such as map projection, on which the dots will be placed. The map in Figure 4.2, with its extreme distortion would be a bad choice for any dot map, and for most others, as well. One needs to think about default values, and whether or not they are good choices for the task at hand, in all contexts of handling data--from fitting curves to the data to fitting maps to the data.

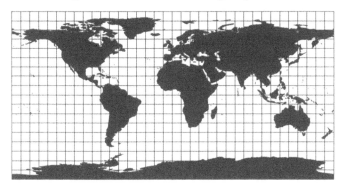

Figure 4.2. Evenly-spaced grid used to draw a map not well-suited to much use--note extreme distortion toward either pole.

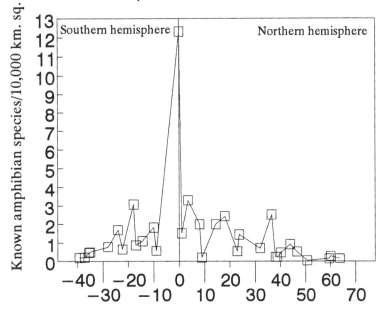

Figure 4.3. Known amphibian species per 10,000 km. sq. by latitude. Equator at 0.

When the data on known amphibian species by country are coupled with country area, it is an easy matter to calculate the density of known amphibian species per country. The fifth column in Table 4.1 shows the results of performing that calculation--dividing the number of species by the area of the country. These numbers are comparable: Ecuador, Colombia, and Zimbabwe have the most dense diversity of known amphibian species; Canada, Finland, and Norway in the northern hemisphere, and New Zealand and Australia in the southern hemisphere, have the least dense diversity of known amphibian species.

Following this lead, latitude was introduced into Table 4.1; when density of amphibian diversity is plotted as a function of latitude, there appears to be some association between the two variables (Figure 4.3). There is considerable variation away from a direct relationship; indeed, there are deserts near the equator where species diversity will drop. In short, there are numerous other factors to be considered too.

Map Overlays

Amphibians are boundary dwellers (Nystuen 1964) between land and fresh water; they bounce back and forth between land and water, anchored in one, and deriving strength from the other, at various stages in their lives. The global distribution of amphibians will also depend on the distribution of land; because the distribution of amphibians is being considered as a function of latitude, it makes sense also to consider the distribution of landmasses as a function of latitude. Figure 4.4 shows buffers, regions surrounding a selected object on an electronic map, surrounding lines of latitude. The evenly-spaced latitude lines, -60, -50, -40, -30, -20, -10, 0, 10, 20, 30, 40, 50, 60 are the midlines of each of these horizontal buffers. Each buffer is about 347 miles wide--a figure obtained by dividing 12,500 (half the circumference of an earth-sphere) by 18 (the number of regions determined by parallels of latitude on a grid with spacing of 10 degrees).

It is possible to use a GIS to find the intersection of countries in the database underlying the GIS with each of these buffers. Figure 4.5 shows buffers 3, 7, and 11 highlighted with their underlying landmasses, selected by a GIS. Notice that this strategy is merely approximate; the U.S.A. for example is counted in buffer 3. Clearly, though, its north-south extent is greater than 10 degrees of latitude so that the fit is not exact. The software assigned an entire country to a strip of latitude based on some central point, such as a centroid.

Figure 4.4.   Buffer zones of ten degrees of latitude; for reference, buffer 1 is the uppermost buffer; buffer 7 is the equatorial buffer; and, buffer 13 is the bottommost buffer. These are superimposed on a Mollweide projection of the world; the midlines of the horizontal buffers are latitude lines (parallels) from 60 degrees north to 60 degrees south.

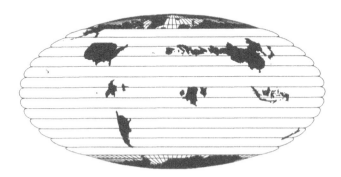

Figure 4.5. Buffers 3, 7, and 11 are highlighted with countries assigned to them by a GIS from the underlying data base. Notice that the fit of country boundaries (curves) to buffers is far from exact; nonetheless, it is useful when coupled with other data.

TABLE 4.2: COUNTRY AREAS BY LATITUDE BUFFERS

Countries by buffer	Total land per buffer	Area by country	Area per buffer	% land by buffer	Lat.
Buffer 1	10998147		**8,998,000**	122.2	60
Canada		3851809			
Denmark		16629			
Estonia		17413			
Finland		130129			
Iceland		39800			
Latvia		24695			
Lithuania		26173			
Norway		125050			
Russia		6592800			
Sweden		173649			
Buffer 2	2823472		**11,460,000**	24.63	50
Austria		32374			
Belgium		11781			
Bylorussia		80200			
Czechoslovakia		49373			
France		210039			
Germany		137727			
Hungary		35920			
Ireland		27137			
Kazakhstan		1049200			
Liechtenstein		62			
Luxembourg		998			
Moldavia		13012			
Mongolia		604200			
Netherlands		15770			
Poland		120725			
Romania		91699			
Switzerland		15941			
Ukraine		233100			
United Kingdom		94214			
Buffer 3	9013274.6		**13,590,000**	66.32	40
Albania		11100			
Andorra		175			
Armenia		11306			
Azerbaijian		33400			
Azores		905			
Bulgaria		42823			

China		3691500		
Cyprus		3572		
Georgia		26911		
Gibraltar		2		
Greece		50944		
Italy		116313		
Japan		143706		
Kirghizia		76642		
Malta		122		
Monaco		0.6		
North Korea		46540		
Portugal		35553		
San Marino		24		
South Korea		38022		
Spain		194885		
Tadzhikistan		54019		
Turkey		301382		
Turkmenistan		186417		
U.S.A.		3675545		
Uzbekistan		172700		
Yugoslavia		98766		
Buffer 4	3838537		**15,320,000**	25.05	30
Afghanistan		250000		
Algeria		919595		
Bahamas		5380		
Bahrain		231		
Bermuda		21		
Bhutan		18200		
Canary Is.		2807		
Egypt		386900		
Iran		636300		
Iraq		167925		
Israel		8019		
Jordan		37738		
Kuwait		6200		
Lebanon		3950		
Libya		679362		
Midway Is.		2		
Morocco		172415		
Nepal		54362		
Pakistan		345753		
Qatar		8500		
Syria		71498		
Tunisia		63379		

Buffer 5	5888282		**16,590,000**	35.49	20
Anguilla		35			
Antigua & Barbuda		171			
Bangladesh		55126			
Belize		8867			
Br. Virgin Is.		59			
Cape Verde		1557			
Cayman Is.		100			
Chad		495800			
Cuba		44218			
Dominica		290			
Dom. Rep.		18816			
Guadeloupe		687			
Guatemala		42042			
Haiti		10714			
Hong Kong		399			
India		1229210			
Jamaica		4232			
Laos		91400			
Mali		478655			
Mauritania		397950			
Mexico		761604			
Myanmar		261790			
Niger		489200			
Oman		82030			
Philippines		115831			
Puerto Rico		3435			
Saudi Arabia		830000			
St. Kitts & Nevis		103			
Taiwan		13885			
Turks & Caicos Is.		166			
U.S. Virgin Islands		133			
U.A.E.		32300			
Vietnam		128402			
W. Sahara		102700			
Yemen		186375			
Buffer 6	3852446		**17,370,000**	22.17	10
Barbados		166			
Benin		43484			
Burkina Faso		105800			
Cameroon		183569			
C. Afr. Rep.		240535			
Costa Rica		19650			
Djibouti		8900			
El Salvador		8260			
Ethiopia		471778			
Gambia		4361			
Ghana		92100			

Grenada	133
Guam	212
Guinea	94962
Guinea–Bissau	13948
Honduras	43277
Ivory Coast	124504
Liberia	43000
Martinque	425
Nicaragua	50200
Nigeria	356669
Panama	29209
Senegal	75955
Sierra Leone	27699
Somalia	246201
Sri Lanka	25330
St. Lucia	238
St. Vincent	150
Sudan	967507
Thailand	198500
Togo	21600
Trinidad	1980
Venezuela	352144

Buffer 7

Buffer 7	3104571		**17,630,000**	17.60	0
Brunei		2226			
Burundi		10747			
Colombia		439737			
Congo		132000			
Ecuador		109483			
Equatorial Guinea		10830			
French Guiana		35100			
Gabon		103347			
Guyana		83000			
Indonesia		753721			
Kenya		224960			
Kiribati		266			
Malaysia		128430			
Maldives		115			
Nauru		8			
Rwanda		10169			
Sao Tome		372			
Seychelles		156			
Singapore		224			
Suriname		63037			
Uganda		91076			
Zaire		905567			

Buffer 8	5157203.1		**17,370,000**	29.69	−10
Angola		481353			
Brazil		3286487			
Comoros		803			
Cook Island		93			
Malawi		45747			
Mayotte		144			
Papua New Guinea		178260			
Peru		496224			
Solomon Islands		11500			
Tanzania		364900			
Tuvalu		9.1			
Western Samoa		1097			
Zambia		290586			

Buffer 9	4803777		**16,590,000**	28.95	−20
Australia		2967909			
Bolivia		424164			
Botswana		231805			
Fiji		7055			
French Polynesia		1550			
Madagascar		226658			
Mauritius		789			
Mozambique		303771			
Namibia		317827			
New Caledonia		7358			
Niue		100			
Paraguay		157048			
Reunion		969			
Tonga		270			
Vanuatu		5700			
Zimbabwe		150804			

Buffer 10	558840		**15,320,000**	3.647	−30
Lesotho		11720			
South Africa		471879			
Swaziland		6705			
Uruguay		68536			

Buffer 11	1468156		**13,590,000**	10.80	−40
Argentina		1072162			
Chile		292258			
New Zealand		103736			

Buffer 12	4618		**11,460,000**	0.040	−50
Falkland Islands		4618			

Buffer 13	0		**8,998,000**	0	−60

Different kinds of software might offer capabilities for finding different sorts of intersections--some might be more exact, but those would sacrifice the easy recognition of shape of the country--it is an easy matter to see that Argentina has been assigned to Buffer 11; if only that part of Argentina actually in Buffer 11 had been assigned to the buffer, one could likely not tell, by shape alone, that indeed that section of land in Buffer 11 belonged to Argentina. There are advantages and drawbacks to every assignment pattern. Indeed, geographer Waldo Tobler proposes a clever idea in his scheme of postal addresses based on latitude and longitude--exact intersections, down to the point level, would become available for finding intersection patterns; then, one would need to be certain to have a scheme to fill in between the points in order to represent a continuous stretch of area.

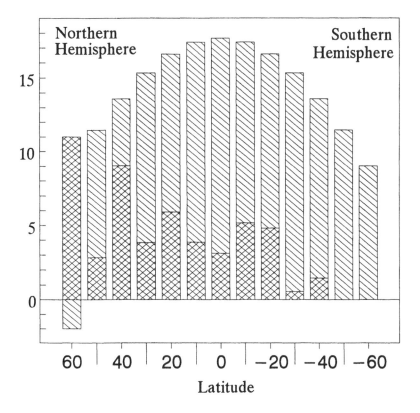

Figure 4.6. Symmetric long bars show the amount of area in each latitudinal buffer; checked areas show an amount of land area contained within that buffer. Because the fit of landmasses to buffers is only approximate (based on country centroids), extreme overfit occurred in the leftmost buffer.

In this sort of map analysis, the curves of political boundaries are "fit" to buffers. Indeed, in the topmost buffer, buffer 1, more land is counted than can actually fit within the theoretical limits of the buffer (see Table 4.2); hence the % of land within the buffer is over 122%! It is because of the assignment pattern of country centroid (and the attached country shape) to buffer that this overshoot occurs.

The GIS calculates the area within each buffer and informs the user which countries are assigned, via country centroid, to each buffer--this assignment is given in Table 4.2. It is then a fairly easy matter to use either the GIS database, or other more familiar databases such as atlases or almanacs, to find the area of each country in each buffer. Entering this data in a spreadsheet makes it easy to calculate the percentage of land in each buffer (Table 4.2). The data in Table 4.2 can then be graphed to get some sort of picture as to the distribution of land within each latitudinal buffer (Figure 4.6); evidently, there is more land in the northern buffers and less in the southern buffers, illustrating clearly why it is that cylindrical map projections are often truncated in a manner that may appear to emphasize the northern hemisphere--there is simply more land, and if as large a map scale as is possible is to be used, and if the focus is on landmasses, then it makes sense not to use valuable map space showing regions of water devoid of landmasses.

When observations about the distribution of landmasses by latitude are coupled with observations about the distribution of amphibians by latitude, the apparently low density of amphibian species in the northern latitudes (from the amphibian data alone) seems even more dramatic-- the latitudes of the earth with the most land space are species-poor.

To investigate the theme of latitude and biodiversity further, one might then wish to introduce data concerning other species/taxa. There is not much data in Table 4.1 concerning Africa. The data sets concerning known plant taxa offer more global coverage. When these data are arranged (Table 4.3) as were those for the amphibian species, patterns appear in the distribution of plant taxa that are similar to those for amphibian species.

Once again, Canada is the country with the lowest density of species and Ecuador the country with the highest diversity. Between these extremes there is variation between the lists; some of it reflects desertification of countries in Africa--the sub-tropical countries of Niger, Mauritania, Chad, Sudan, and Mali, as well as the large, nearly sub-tropical, countries of Libya, Algeria, and Egypt exhibit low plant taxa diversity. Some of the other differences between the tables are due to different requirements from the environment for amphibians and plants. Nonetheless, the same sort of general pattern appears when known plant taxa are plotted as a function of latitude (Figure 4.7), but the association seems not as clear as for amphibians.

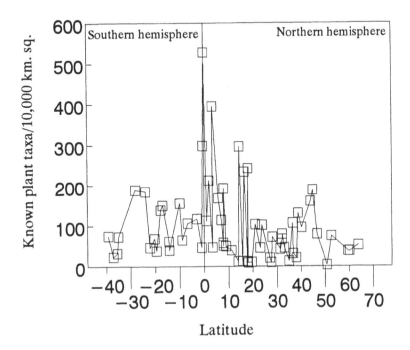

Figure 4.7. Known plant taxa per 10,000 km. sq. arranged by latitude.

Indeed, when the known plant taxa data are plotted as a dot map (Figure 4.8), there are evident clusters in equatorial countries except where there has been substantial desertification; there are also low numbers of different taxa in northern high latitudes, perhaps suggesting that diversity in plants is a long process--longer than the time since the most recent episodes of continental glaciation at these latitudes. Plants cannot migrate away from advancing glaciers.

One might wonder what a dot map based on 20 year old or 50 year old data sets of this sort would show in sub-Saharan Africa--presumably a situation more like what is currently displayed in Ecuador and Colombia with great diversity of taxa. It is for this reason that monitoring change in diversity is important. Indeed, is there a threshold value that can be captured, beyond which extinction of species will occur? Chaos theory--Feigenbaum's graphical analysis--offers one method to find a threshold of irreversibility and to alter it.

TABLE 4.3: KNOWN PLANT TAXA
AND OTHER DATA
(Sources: World Resources Institute Data Base;
Goode's World Atlas)

Latitude	Country	Known Plant Taxa	Area km. sq.	Known plant Taxa per 10,000 km. sq.
50.72	Canada	3220	9,976,000	3.2277465918
18.03	Niger	1178	1,267,000	9.2975532755
27.63	Libya	1700	1,760,000	9.6590909091
19.63	Mauritania	1100	1,026,000	10.7212475634
17.8	Chad	1600	1,284,000	12.46105919
14	Sudan	3200	2,506,000	12.7693535515
15.75	Mali	1600	1,240,000	12.9032258065
34.97	Algeria	3145	2,382,000	13.2031905961
26.97	Egypt	2085	1,001,000	20.829170829
38	United States	20000	9,373,000	21.337885416
−37.17	Australia	18000	7,687,000	23.416157148
36.75	China	30000	9,561,000	31.377470976
−35.5	Argentina	9000	2,767,000	32.526201662
60.17	Sweden	1700	450,000	37.777777778
−19.5	Namibia	3159	824,000	38.337378641
59.58	Finland	1300	338,000	38.461538462
11.17	Burkina Faso	1096	274,000	40
−14.25	Angola	5000	1,247,000	40.096230954
31.25	Iran, Islamic Rep	7000	1,648,000	42.475728155
23	India	15000	3,288,000	45.620437956
33	Afghanistan	3000	652,000	46.012269939
−22.17	Botswana	2700	582,000	46.391752577
−1	Zaire	11000	2,345,000	46.908315565
3.47	Somalia	3000	638,000	47.021943574
8.95	Nigeria	4614	924,000	49.935064935
7.88	Ethiopia	6283	1,222,000	51.415711948
63.8	Norway	1700	324,000	52.469135802
7.83	Central African Rep	3600	623,000	57.784911717
−14.38	Zambia	4600	753,000	61.088977424
−9	Brazil	55000	8,512,000	64.614661654
32	Iraq	2937	438,000	67.054794521
−20.25	Mozambique	5500	802,000	68.578553616
28	Pakistan	5750	796,000	72.236180905
−35	Chile	5500	757,000	72.655217966
−39.23	New Zealand	2000	269,000	74.349442379
52.62	Poland	2350	313,000	75.079872204
46.65	France	4375	552,000	79.257246377
32	Morocco	3550	447,000	79.418344519
40.25	Spain	4825	505,000	95.544554455
23.75	Mexico	20000	1,958,000	102.145045965
21	Myanmar	7000	677,000	103.397341211
−6.8	Tanzania	10000	945,000	105.82010582
36.5	Japan	4022	378,000	106.402116402
1	Kenya	6500	580,000	112.068965517
7.12	Cote d'Ivoire	3660	322,000	113.664596273
−3	Congo	4000	342,000	116.959064327
38.75	Turkey	10150	779,000	130.29525032
−17.83	Zimbabwe	5428	391,000	138.82352941
−17	Bolivia	16500	1,099,000	150.13648772

−10	Peru	20000	1,285,000	155.64202335
43.97	Italy	4825	301,000	160.29900332
5.8	Cameroon	8000	475,000	168.42105263
−24	Paraguay	7500	407,000	184.27518428
−28	South Africa	23000	1,221,000	188.37018837
44.8	Yugoslavia	4825	256,000	188.4765625
8	Venezuela	17500	912,000	191.88596491
2	Uganda	5000	236,000	211.86440678
16.5	Thailand	12000	513,000	233.91812865
18	Viet Nam	8000	330,000	242.42424242
14.42	Philippines	8900	300,000	296.66666667
−0.5	Gabon	8000	268,000	298.50746269
3.5	Colombia	45000	1,139,000	395.0834065
0	Ecuador	15000	284,000	528.16901408

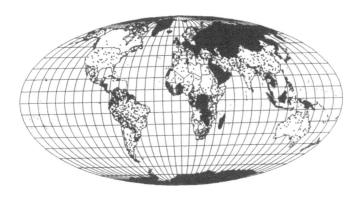

Figure 4.8. Distribution of known plant taxa for selected countries. One dot represents about 300 different taxa. Mollweide equal-area projection is used as base map. Land masses for which there is no data in this set are shaded.

Feigenbaum's graphical analysis

Feigenbaum's graphical analysis (Feigenbaum, 1980) is a tool from mathematical chaos theory that offers a strategy to understand how small geometric changes lead to large geometric differences. This form of analysis rests on an ordering of events that is not necessarily temporal, but in which the output of one stage serves as the input for the next stage. In Figure 4.9, the line y=x is used as an axis in which the output of one stage becomes the input of the next. An input of x, in

Figure 4.9, leads to an output of y, which is then used (after shifting horizontally to the line y=x) as an input to produce an output of y'; then y' is used as an input to generate y", and so forth. Instead of reading input values from the x-axis, they are read from y=x so that the resulting geometric pattern, in this case, is a rising staircase. The initial value, (x,0), is called the "seed" value of the analysis. The geometric pattern, forced on the trajectory of the seed by the relative positions of the curve and the line y=x, is called the orbit of the seed value. In this case, seed values to the right of (P,0) (but not past the next intersection of the curve with y=x) all generate ascending staircases, which may subsequently exhibit even greater geometric complexity; those to the left of (P,0) (but not to the left of the previous intersection of the curve with y=x) all generate descending staircases.

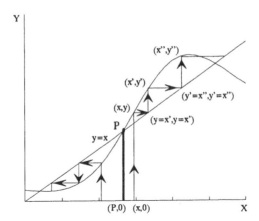

Figure 4.9. Feigenbaum's graphical analysis applied to a curve; to the right of P orbits are ascending staircases; to the left they are descending.

If the descending situation indicates a favorable geometric dynamic--one that is under control--then the point P suggests a threshold of irreversibility; beyond it, the geometric process takes off in an undesired direction (Arlinghaus, Nystuen, and Woldenberg, 1992). However, because P is found as an intersection of a curve and a line, a slowing of the increase, anywhere to the left of P, means that the threshold is shifted further to the right--or that its attainment is delayed and may be delayed indefinitely as long as intervention to the left of P, to control the increase in geometric process, continues to prevent the intersection of the curve and the line y=x. Abstract tools such as this one offer a great deal of promise in suggesting directions for theoretical research--the critical component underlying any form of analysis of data.

Summary

Generally, these associations suggests that countries closer to the poles have a less dense diversity of species than do those near the equator; there may be a variety of reasons for this--certainly one might expect that lush tropical rainforests would support a greater diversity of species than would other locales. Indeed, the most recent episode of continental glaciation may have made the recently glaciated regions of northern latitudes even less densely endowed with diverse species than one would expect on the basis of latitude, and climatic factors, alone. Islands may have unusual patterns of diversity based on their remoteness from mainland sources for new species and taxa. The data here is inconclusive, but suggestive of various ideas. It is important, though, to note such speculation because it could be used, in a positive way by policy makers--indeed, even in international trade between species-rich and species-poor countries, and it can guide decisions on how to accumulate data sets stemming from the baseline set. Creativity is critical in making adjustments for dealing with data--in handling problems of missing data as well as in ensuring proper proportions.

References

1. Arlinghaus, S. L., Nystuen, J. D., Woldenberg, M. J. An application of graphical analysis to semidesert soils. *Geographical Review*, 1992, Vol. 82, No. 3, pp. 244-252.
2. Feigenbaum, M. J. 1980. Universal behavior in non-linear systems. *Los Alamos Science*, summer, 4-27.
3. Kates, R. W. and Burton, I. *Geography, Resources, and Environment*, 1986, Chicago, University of Chicago Press.
4. Nystuen, J. D. Effects of boundary shape and the concept of local convexity. 1966. *Papers*, Michigan Inter-University Community of Mathematical Geographers, 10:3-24, Ann Arbor.
5. Tobler, W. R. Personal communication. 1993.

CHAPTER 5

SOILS AND FORESTRY DATA ANALYSIS

ANALYTICAL TECHNIQUES/TOOLS USED:

Simple cubic spline curve fitting
Interpolation using a cubic spline
Feigenbaum's graphical analysis

DATA TYPE: HYPOTHETICAL

Overview of Data

In the data base of the World Resources Institute, there are many variables that deal with forestry and soils: from deforestation data expressed in percentages and thousands of hectares for closed forests and total forests to soil degradation data based on 22 different indicators (as for example wind erosion and waterlogging). Instead of choosing one or the other of these variables, we review instead, in detail, a different way of looking at data of this sort (an expansion of the article listed in the references by Arlinghaus, Nystuen, and Woldenberg). We do so using hypothetical data so that we might expose the reader, in detail, to the mechanics of two complex analytical tools: cubic spline curve fitting and Feigenbaum's graphical analysis. Then, we indicate how these might be interpreted in the context of a soils/forestry setting.

Simple cubic spline curve fitting

The cartographer's spline can be imitated using mathematics: hence the name "spline." When the mathematical spline is composed of pieces of polynomials of degree three, cubics, the procedure is cubic spline fitting. Roughly, the idea is to fit pieces of cubic curves between a finite set of sample data points: one cubic is fit between two adjacent sample points and another cubic is fit between another pair of adjacent sample points. In this manner, one can continue piecing together a curve between a finite set of pairs of adjacent points. Because the curve is fit between points, and never extended beyond any sample point, it is a fit that is bounded and is useful only for interpolation--not for extrapolation.

BLACK BOX SUMMARY
see Introduction
SIMPLE CUBIC SPLINE CURVE FITTING
Fit a cubic, S(x), of the form
to each interval
between a finite set of given evenly-spaced
sample points, one unit apart,
$(x_1, y_1),...,(x_n, y_n)$.
The spline, S(x), composed of the following n equations

$$a_1(x-x_1)^3 + b_1(x-x_1)^2 + c_1(x-x_1) + d_1$$
$$a_2(x-x_2)^3 + b_2(x-x_2)^2 + c_2(x-x_2) + d_2$$
$$........$$
$$a_{n-1}(x-x_{n-1})^3 + b_{n-1}(x-x_{n-1})^2 + c_{n-1}(x-x_{n-1}) + d_{n-1}$$

that will fit the sample points has coefficients given by:
$$a_i = (M_{i+1} - M_i)/6$$
$$b_i = M_i/2$$
$$c_i = (y_{i+1} - y_i) - ((M_{i+1} + 2M_i)/6)$$
$$d_i = y_i$$
where $i=1,2,...,n-1$
and where the M_i are determined as solutions
to the following matrix equation

$$
\begin{bmatrix}
1 & 0 & 0 & 0 & ... & 0 & 0 & 0 \\
1 & 4 & 1 & 0 & ... & 0 & 0 & 0 \\
0 & 1 & 4 & 1 & ... & 0 & 0 & 0 \\
 & & & ... & & & & \\
0 & 0 & 0 & 0 & ... & 1 & 4 & 1 \\
0 & 0 & 0 & 0 & ... & 0 & 0 & 1
\end{bmatrix}
\begin{bmatrix}
M_1 \\
M_2 \\
M_3 \\
.... \\
M_{n-1} \\
M_n
\end{bmatrix}
= 6
\begin{bmatrix}
0 \\
y_1 - 2y_2 + y_3 \\
y_2 - 2y_3 + y_4 \\
.... \\
y_{n-2} - 2y_{n-1} + y_n \\
0
\end{bmatrix}
$$

where the Ms are a column matrix (vector) that is multiplied on the left
by an (nxn) matrix of constants; this product is equal to 6 times a
column matrix of constants whose values can be determined from the
second coordinates of the given sample points. Thus, to solve for the
column of Ms, it is necessary to find the inverse of the (nxn) matrix of
constants on the far left.

The fit should be such that the curve is smooth and continuous throughout its bounded interval, and so that at the sample points, where the curve is spliced together, a line tangent to the curve has the same slope whether the line is tangent to the curve determined by a cubic to the left of the sample point or whether the line is tangent to the curve determined by a cubic to the right of the sample point. The curve is differentiable.

The Black Box Summary above gives the highlights of cubic spline curve fitting; an example will be worked using six sample points.

Cubic spline fit to six hypothetically-chosen sample points

Consider a set of six points, chosen simply to illustrate the relatively complicated process of fitting five pieces of cubic polynomials between the adjacent points. Figure 5.1 shows the beginning of the analysis. In it, the coordinates of the sample points, spaced one unit apart, are given by:

$$(x_1, y_1) = (1, 1.25)$$
$$(x_2, y_2) = (2, 1.75)$$
$$(x_3, y_3) = (3, 3)$$
$$(x_4, y_4) = (4, 2.5)$$
$$(x_5, y_5) = (5, 2)$$
$$(x_6, y_6) = (6, 1.75)$$

This designation of sample points is followed by rewriting the matrix equation from the Black Box above, substituting the values of the second coordinates, as appropriate, from the sample points.

$$0$$
$$y_1 - 2y_2 + y_3 = 1.25 - 2*1.75 + 3 = 0.75$$
$$y_2 - 2y_3 + y_4 = 1.75 - 2*3 + 2.5 = -1.75$$
$$y_3 - 2y_4 + y_5 = 3 - 2*2.5 + 2 = 0$$
$$y_4 - 2y_5 + y_6 = 2.5 - 2*2 + 1.75 = 0.25$$
$$0 .$$

Thus, the Black Box matrix equation appears as a 6 x 6 matrix of prescribed constants on the left, times $M_1...M_6$ on the left. This is set equal to 6 times the values of the column matrix (vector) found just above. To solve this equation, find the inverse of this matrix. The matrix is written in the second to last part of Figure 5.1. Its inverse is found using a spreadsheet. The inverse is written in the last part of Figure 5.1.

CUBIC SPLINE INTERPOLATION
EXAMPLE- - CHOOSE SIX POINTS
$(x1, y1) = (1, 1.25)$
$(x2, y2) = (2, 1.75)$
$(x3, y3) = (3, 3)$
$(x4, y4) = (4, 2.5)$
$(x5, y5) = (5, 2)$
$(x6, y6) = (6, 1.75)$

Equation from Black Box:

1	0	0	0	0	0		M1		0
1	4	1	0	0	0		M2		0.75
0	1	4	1	0	0	times	M3	= 6 times	−1.75
0	0	1	4	1	0		M4		0
0	0	0	1	4	1		M5		0.25
0	0	0	0	0	1		M6		0

Solve the matrix equation for M1...M6:

M1		1	0	0	0	0	0			0
M2		−0.26	0.267	−0.07	0.019	−0.00	0.004			0.75
M3	= 6 times	0.071	−0.07	0.287	−0.07	0.019	−0.01	times		−1.75
M4		−0.01	0.019	−0.07	0.287	−0.07	0.071			0
M5		0.004	−0.00	0.019	−0.07	0.267	−0.26			0.25
M6		0	0	0	0	0	1			0

		0		0	
		0.325		1.952	
=	6 times	−0.55	=	−3.30	
		0.130		0.782	
		0.029		0.179	
		0		0	

Matrix used:

1	0	0	0	0	0
1	4	1	0	0	0
0	1	4	1	0	0
0	0	1	4	1	0
0	0	0	1	4	1
0	0	0	0	0	1

Inverse used (calculated in spreadsheet):

1	0	0	0	0	0
−0.26	0.267	−0.07	0.019	−0.00	0.004
0.071	−0.07	0.287	−0.07	0.019	−0.01
−0.01	0.019	−0.07	0.287	−0.07	0.071
0.004	−0.00	0.019	−0.07	0.267	−0.26
0	0	0	0	0	1

Figure 5.1. First stages in an example of fitting a cubic spline to six points--finding the Ms.

Find coefficients, a's, b's, c's, and d's as in black box

	a1	a2	a3	a4	a5
0 M1	0.32535				
1.952 M2		−0.8767			
−3.30 M3			0.68181		
0.782 M4				−0.1004	
0.179 M5					−0.0299
0 M6					

	b1	b2	b3	b4	b5
0 M1	0				
1.952 M2		0.97607			
−3.30 M3			−1.6543		
0.782 M4				0.39114	
0.179 M5					0.08971
0 M6					

	c1	c2	c3	c4	c5
0 M1	0.17464				
1.952 M2		1.15071			
−3.30 M3			0.47248		
0.782 M4				−0.7906	
0.179 M5					−0.3098
0 M6					

	d1	d2	d3	d4	d5
0 M1	1.25				
1.952 M2		1.75			
−3.30 M3			3		
0.782 M4				2.5	
0.179 M5					2
0 M6					

a's	b's	c's	d's
0.32535	0	0.17464	1.25
−0.8767	0.97607	1.15071	1.75
0.68181	−1.6543	0.47248	3
−0.1004	0.39114	−0.7906	2.5
−0.0299	0.08971	−0.3098	2

Equation of cubic spline to fit the six sample points.
Five equations to fit curve pieces between the six sample points.

$$S(x) = \begin{array}{llll} 0.325\ (x-1)^3 +0 & (x-1)^2 +0.174641(x-1) & +1.25 \\ -0.87\ (x-2)^3 +0.97607\ (x-2)^2 +1.15071\ (x-2) & +1.75 \\ 0.681\ (x-3)^3 -1.6543\ (x-3)^2 +0.47248\ (x-3) & +3 \\ -0.10\ (x-4)^3 +0.39114\ (x-4)^2 -0.79066\ (x-4) & +2.5 \\ -0.02\ (x-5)^3 +0.08971\ (x-5)^2 -0.3098\ (x-5) & +2 \end{array}$$

Figure 5.2. Second stage of cubic spline fitting--finding a, b, c, d, for each curve piece.

Then, to solve the top matrix equation in Figure 5.1 for the column matrix containing $M_1...M_6$ it is necessary to isolate the column of Ms on the left. Thus, multiply both sides of the first matrix equation *on the left* by the inverse of the matrix (shown at the bottom of Figure 5.1). Then simplify, obtaining, as shown in Figure 5.1, the following values for the Ms, respectively: 0, 1.952, -3.30, 0.782, 0.179, and 0. These values will be used in the next stage of the analysis.

Figure 5.2 uses the values for the six Ms found in Figure 5.1 to find values for the constants, a, b, c, and d that will appear in each of the five pieces of cubic curve fit between the sequence of six points. The letters M_1 to M_6 are placed in a column next to the constants for reference in substituting values. Using the formula in the black box, calculate values for the constant a (results in Figure 5.2):

$$a_1 = (M_2 - M_1)/6 = (1.952-0)/6 = 0.32535$$
$$a_2 = (M_3 - M_2)/6 = (-3.30-1.952)/6 = -0.8767$$
$$a_3 = (M_4 - M_3)/6 = (0.782+3.30)/6 = 0.0.68181$$
$$a_4 = (M_5 - M_4)/6 = (0.179-0.782)/6 = -0.1004$$
$$a_5 = (M_6 - M_5)/6 = (0-0.179)/6 = -0.0299$$

The answers were calculated in a spreadsheet using more decimal places in the input values than shown here. In a similar way, the values for the various constants, b, may be calculated.

$$b_1 = M_1/2 = 0/2 = 0$$
$$b_2 = M_2/2 = 1.952/2 = 0.97607$$
$$b_3 = M_3/2 = -3.30/2 = -1.6543$$
$$b_4 = M_4/2 = 0.782/2 = 0.39114$$
$$b_5 = M_5/2 = 0.179/2 = 0.08971$$

The values for c also follow directly.

$$c_1=(y_2 - y_1)-((M_2+2*M_1)/6)=(1.75-1.25)-(1.952+2*0)/6)=0.17464$$
$$c_2=(y_3 - y_2)-((M_3+2*M_2)/6)=(3-1.75)-(-3.30+2*1.952)/6)=1.15071$$
$$c_3=(y_4 - y_3)-((M_4+2*M_3)/6)=(2.5-3)-(0.782-2*3.30)/6)=0.47248$$
$$c_4=(y_5 - y_4)-((M_5+2*M_4)/6)=(2-2.5)-(0.179+2*0.782)/6)=-0.7906$$
$$c_5=(y_6 - y_5)-((M_6+2*M_5)/6)=(0-2)-(0+2*0.179)/6)=-0.3098$$

The values for d also follow.

$$d_1=y_1=1.25$$
$$d_2=y_2=1.75$$
$$d_3=y_3=3$$
$$d_4=y_4=2.5$$
$$d_5=y_5=2$$

These values are summarized in Figure 5.2 in a four by four matrix. Then each is used to generate five equations to fit curve pieces between the six sample points. These equations are shown in the bottom part of Figure 5.2. The function S(x) is thus defined on a split domain and has the following values.

In the interval between (1, 1.25) and (2, 1.75) the cubic found was
$$0.325(x-1)^3+0(x-1)^2+0.17464(x-1)+1.25;$$
in the interval between (2, 1.75) and (3, 3) the cubic found was
$$-0.87(x-2)^3+0.97607(x-2)^2+1.15071(x-2)+1.75;$$
in the interval between (3, 3) and (4, 2.5) the cubic found was
$$0.681(x-3)^3-1.6543(x-3)^2+0.47248(x-3)+3;$$
in the interval between (4, 2.5) and (5, 2) the cubic found was
$$-0.10(x-4)^3+0.39114(x-4)^2-0.79066(x-4)+2.5;$$
in the interval between (5, 2) and (6, 1.75) the cubic found was
$$-0.02(x-5)^3+0.08971(x-5)^2-0.3098(x-5)+2.$$

Then, each piece of cubic may be used to find other points between the two endpoints--to interpolate between the endpoints.

Figure 5.3 shows values calculated at tenths of a unit using each equation over the domain in which it is defined. Figure 5.4 shows a graph derived from the values of Figure 5.3, and Figure 5.5 shows the equation used to calculate each of the values in Figure 5.4--it is important to recognize that different equations must be used in different intervals. The curve in Figure 5.4 looks perfectly smooth, as it should; but, it is calculated using five different equations. That characteristic is the hallmark of fitting a curve using a spline--mathematical or other.

Plot point by point; tenths of a unit between sample points.

x value	spline fit	
1	1.25	
1.1	1.26778	
1.2	1.28753	
1.3	1.31117	
1.4	1.34067	
1.5	1.37799	
1.6	1.42506	
1.7	1.48384	
1.8	1.55629	
1.9	1.64436	
2	1.75	same value using equations from above and below
2.1	1.87395	
2.2	2.01217	
2.3	2.15938	
2.4	2.31034	
2.5	2.45977	
2.6	2.60243	
2.7	2.73304	
2.8	2.84634	
2.9	2.93708	
3	3	same value using equations from above and below
3.1	3.03138	
3.2	3.03378	
3.3	3.01126	
3.4	2.96794	
3.5	2.90789	
3.6	2.83521	
3.7	2.75399	
3.8	2.66832	
3.9	2.58230	
4	2.5	same value using equations from above and below
4.1	2.42474	
4.2	2.35671	
4.3	2.29529	
4.4	2.23989	
4.5	2.18990	
4.6	2.14472	
4.7	2.10374	
4.8	2.06638	
4.9	2.03201	
5	2	same value using equations from above and below
5.1	1.96988	
5.2	1.94138	
5.3	1.91432	
5.4	1.88852	
5.5	1.86379	
5.6	1.83995	
5.7	1.81684	
5.8	1.79426	
5.9	1.77204	
6	1.75001	

Figure 5.3. Values calculated using the equations of Figure 5.2 for the six sample points and for intervening values spaced at tenths of a unit.

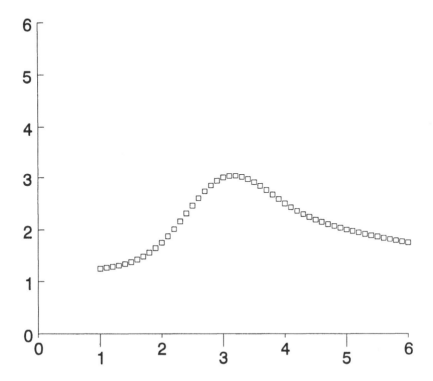

Figure 5.4. Graph of values in Figure 5.3. A cubic spline fit to the six sample points of Figure 5.1.

Feigenbaum's graphical analysis

Graphical analysis is a tool from mathematical chaos theory used to study geometric dynamics (Feigenbaum 1980; Gleick 1987; Devaney and Keen 1989). It rests on an ordering of events in which the output of one stage serves as input for the next.   To see how this works, geometrically, consider the curve in Figure 5.6. This curve is shaped something like the letter S; the line y=x has been superimposed on it. Choose an initial value of (x,0) marked on the x-axis (a "seed" value). To understand the geometric dynamics of this seed, relative to the given curve, repeat the input/output process. That is, if the equation of the curve is y=f(x), then f carries an input of x to an output of y--the point (x,y) on the curve. Repeat the process--that is, use the output, y, as the next input--as an "x"--labelled x' in Figure 5.6.   To do so, move horizontally to the line y=x; now the y becomes an x. Then repeat the process. Send x' to the curve, slide across the horizontal to y=x, and so on. The result is a staircase of values with directions shown by arrows.

```
  1  0.325358*(C103−1)^3+0*(C103−1)^2+0.174641*(C103−1)+1.25
1.1  0.325358*(C104−1)^3+0*(C104−1)^2+0.174641*(C104−1)+1.25
1.2  0.325358*(C105−1)^3+0*(C105−1)^2+0.174641*(C105−1)+1.25
1.3  0.325358*(C106−1)^3+0*(C106−1)^2+0.174641*(C106−1)+1.25
1.4  0.325358*(C107−1)^3+0*(C107−1)^2+0.174641*(C107−1)+1.25
1.5  0.325358*(C108−1)^3+0*(C108−1)^2+0.174641*(C108−1)+1.25
1.6  0.325358*(C109−1)^3+0*(C109−1)^2+0.174641*(C109−1)+1.25
1.7  0.325358*(C110−1)^3+0*(C110−1)^2+0.174641*(C110−1)+1.25
1.8  0.325358*(C111−1)^3+0*(C111−1)^2+0.174641*(C111−1)+1.25
1.9  0.325358*(C112−1)^3+0*(C112−1)^2+0.174641*(C112−1)+1.25
  2  −0.87679*(C113−2)^3+0.976076*(C113−2)^2+1.150717*(C113−2)+1.75
2.1  −0.87679*(C114−2)^3+0.976076*(C114−2)^2+1.150717*(C114−2)+1.75
2.2  −0.87679*(C115−2)^3+0.976076*(C115−2)^2+1.150717*(C115−2)+1.75
2.3  −0.87679*(C116−2)^3+0.976076*(C116−2)^2+1.150717*(C116−2)+1.75
2.4  −0.87679*(C117−2)^3+0.976076*(C117−2)^2+1.150717*(C117−2)+1.75
2.5  −0.87679*(C118−2)^3+0.976076*(C118−2)^2+1.150717*(C118−2)+1.75
2.6  −0.87679*(C119−2)^3+0.976076*(C119−2)^2+1.150717*(C119−2)+1.75
2.7  −0.87679*(C120−2)^3+0.976076*(C120−2)^2+1.150717*(C120−2)+1.75
2.8  −0.87679*(C121−2)^3+0.976076*(C121−2)^2+1.150717*(C121−2)+1.75
2.9  −0.87679*(C122−2)^3+0.976076*(C122−2)^2+1.150717*(C122−2)+1.75
  3  0.681818*(C123−3)^3−1.6543*(C123−3)^2+0.472488*(C123−3)+3
3.1  0.681818*(C124−3)^3−1.6543*(C124−3)^2+0.472488*(C124−3)+3
3.2  0.681818*(C125−3)^3−1.6543*(C125−3)^2+0.472488*(C125−3)+3
3.3  0.681818*(C126−3)^3−1.6543*(C126−3)^2+0.472488*(C126−3)+3
3.4  0.681818*(C127−3)^3−1.6543*(C127−3)^2+0.472488*(C127−3)+3
3.5  0.681818*(C128−3)^3−1.6543*(C128−3)^2+0.472488*(C128−3)+3
3.6  0.681818*(C129−3)^3−1.6543*(C129−3)^2+0.472488*(C129−3)+3
3.7  0.681818*(C130−3)^3−1.6543*(C130−3)^2+0.472488*(C130−3)+3
3.8  0.681818*(C131−3)^3−1.6543*(C131−3)^2+0.472488*(C131−3)+3
3.9  0.681818*(C132−3)^3−1.6543*(C132−3)^2+0.472488*(C132−3)+3
  4  −0.10044*(C133−4)^3+0.391148*(C133−4)^2−0.79066*(C133−4)+2.5
4.1  −0.10044*(C134−4)^3+0.391148*(C134−4)^2−0.79066*(C134−4)+2.5
4.2  −0.10044*(C135−4)^3+0.391148*(C135−4)^2−0.79066*(C135−4)+2.5
4.3  −0.10044*(C136−4)^3+0.391148*(C136−4)^2−0.79066*(C136−4)+2.5
4.4  −0.10044*(C137−4)^3+0.391148*(C137−4)^2−0.79066*(C137−4)+2.5
4.5  −0.10044*(C138−4)^3+0.391148*(C138−4)^2−0.79066*(C138−4)+2.5
4.6  −0.10044*(C139−4)^3+0.391148*(C139−4)^2−0.79066*(C139−4)+2.5
4.7  −0.10044*(C140−4)^3+0.391148*(C140−4)^2−0.79066*(C140−4)+2.5
4.8  −0.10044*(C141−4)^3+0.391148*(C141−4)^2−0.79066*(C141−4)+2.5
4.9  −0.10044*(C142−4)^3+0.391148*(C142−4)^2−0.79066*(C142−4)+2.5
  5  −0.0299*(C143−5)^3+0.089712*(C143−5)^2−0.3098*(C143−5)+2
5.1  −0.0299*(C144−5)^3+0.089712*(C144−5)^2−0.3098*(C144−5)+2
5.2  −0.0299*(C145−5)^3+0.089712*(C145−5)^2−0.3098*(C145−5)+2
5.3  −0.0299*(C146−5)^3+0.089712*(C146−5)^2−0.3098*(C146−5)+2
5.4  −0.0299*(C147−5)^3+0.089712*(C147−5)^2−0.3098*(C147−5)+2
5.5  −0.0299*(C148−5)^3+0.089712*(C148−5)^2−0.3098*(C148−5)+2
5.6  −0.0299*(C149−5)^3+0.089712*(C149−5)^2−0.3098*(C149−5)+2
5.7  −0.0299*(C150−5)^3+0.089712*(C150−5)^2−0.3098*(C150−5)+2
5.8  −0.0299*(C151−5)^3+0.089712*(C151−5)^2−0.3098*(C151−5)+2
5.9  −0.0299*(C152−5)^3+0.089712*(C152−5)^2−0.3098*(C152−5)+2
  6  −0.0299*(C153−5)^3+0.089712*(C153−5)^2−0.3098*(C153−5)+2
```

Figure 5.5. Cubic equations used to calculate the values in Figure 5.3.

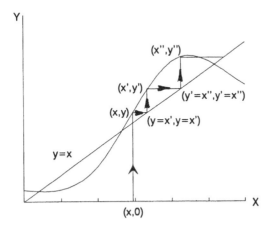

Figure 5.6. Graphical analysis. The output of one stage becomes the input of the next stage--as geometric feedback. From a figure that appeared originally in the *Geographical Review*, Arlinghaus, Nystuen, and Woldenberg reference. Reprinted here with permission of the American Geographical Society.

As the relative position of the curve, whatever it might be, and the line y=x shifts, so too does the pattern of geometric feedback. We can superimpose the line y=x on the cubic spline of Figure 5.4, and on variants of it generated by slight shifts in the underlying set of six sample points (Figure 5.7).

Figure 5.7a.   Spline is fit to sample points:
(1, 1.25), (2, 1.75), (3,3), (4, 2.5), (5,2), (6,1.75).

Figure 5.7b.   Spline is fit to sample points:
(1, 1.25), (2, 1.75), (3,4), (4, 3.5), (5,3), (6,2.75).

Figure 5.7c.   Spline is fit to sample points:
(1, 1.25), (2, 1.75), (3,4), (4, 3), (5,2.), (6,1.75).

Figure 5.7d.   Spline is fit to sample points:
(1, 1.25), (2, 1.75), (3,4), (4, 2.25), (5,2.), (6,1.75).

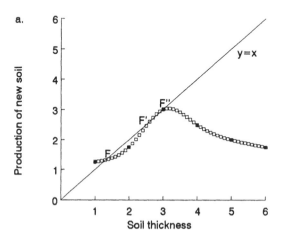

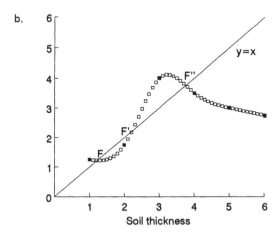

Figure 5.7a and b. Two cubic splines. From a figure that appeared originally in the *Geographical Review*, Arlinghaus, Nystuen, and Woldenberg reference. Reprinted here with permission of the American Geographical Society.

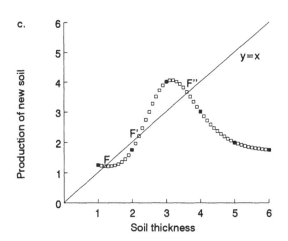

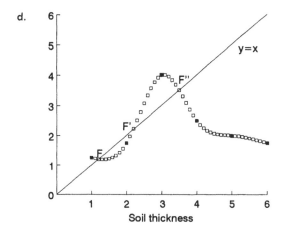

Figure 5.7c and d. Two more cubic splines. From a figure that appeared originally in the *Geographical Review*, Arlinghaus, Nystuen, and Woldenberg reference. Reprinted here with permission of the American Geographical Society.

The reader interested in learning to fit splines to sample points might try using the example explained in detail earlier in this chapter as a guide to fit splines to these other sets of sample points and then check the graphed results of the fit against the corresponding graphs in Figure 5.7. The curves in Figure 5.7 are generally similar in various ways--the small variations in the set of sample points is designed to produce different intersections patterns with y=x which will then lead to vastly different trajectories a seed value might take in graphical analysis.

Figure 5.8 shows these trajectories; the top graph on each page shows the general path, or orbit, of the seed value x=2.5 relative to the curve--the orbit describes the nature of the geometric feedback. The bottom graph shows an enlargement of the detail of the top graph.

In Figure 5.8a
feedback is along a descending staircase;
in Figure 5.8b
feedback wraps around a square spiral;
in Figure 5.8c
feedback bounces in and out of a square tunnel;
in Figure 5.8d
feedback takes on a new form.

Theoretical application to semidesert soil production

Semidesert soils are generally thin and vulnerable to changes in biomass. They typically appear in a desert with some rainfall, some vegetation, and therefore, some soil thickness. This modest soil cover reflects an equilibrium between the erosion rate and the soil-production rate. An increase or decrease in biomass corresponds with changes in soil thickness. The organic semidesert soil system feeds on itself--feedback is natural and critical in maintaining and promoting healthy soil.

Desertification along desert edges might come from human abuse of this fragile system. The biological potential of the soil is subjected to a downward spiral from which the soil cannot recover and collapse of the soil, as a system supporting biological activity becomes imminent. This situation can be reversed if the destruction is not too far underway. When there is plant cover sufficient to protect the soil from the splattering of raindrops, soil thickness increases from decayed vegetation. Humans might introduce plant cover, not to exceed the water capacity of the desert, as a means for recovery. Determining the appropriate point at which to intervene is crucial, if humans are to aid in this cause.

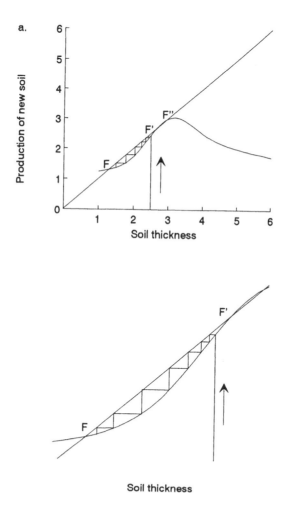

Figure 5.8a. Graphical analysis of a cubic spline; x=2.5. From a figure that appeared originally in the *Geographical Review*, Arlinghaus, Nystuen, and Woldenberg reference. Reprinted here with permission of the American Geographical Society.

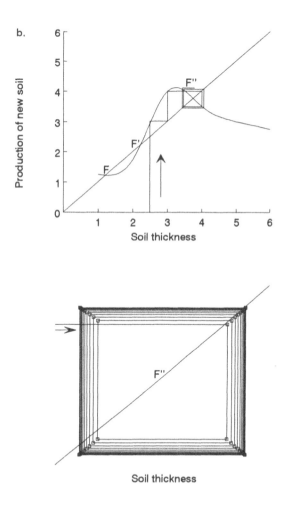

Figure 5.8b. Graphical analysis of a cubic spline; x=2.5. From a figure that appeared originally in the *Geographical Review*, Arlinghaus, Nystuen, and Woldenberg reference. Reprinted here with permission of the American Geographical Society.

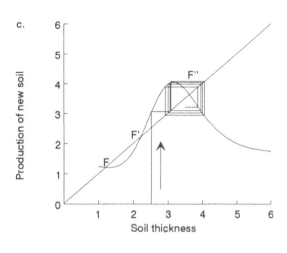

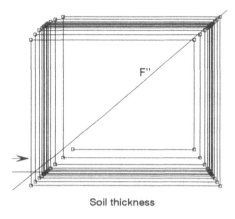

Figure 5.8c. Graphical analysis of a cubic spline; x=2.5. From a figure that appeared originally in the *Geographical Review*, Arlinghaus, Nystuen, and Woldenberg reference. Reprinted here with permission of the American Geographical Society.

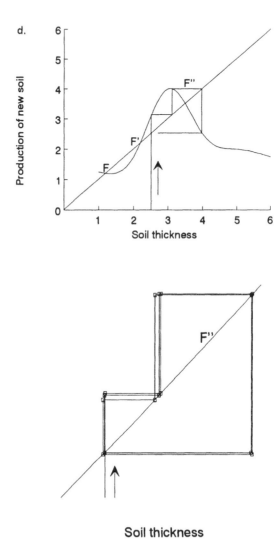

Figure 5.8d. Graphical analysis of a cubic spline; x=2.5. From a figure that appeared originally in the *Geographical Review*, Arlinghaus, Nystuen, and Woldenberg reference. Reprinted here with permission of the American Geographical Society.

Any curve in Figure 5.7 describes the production pattern of semidesert soil. To optimize human aid in soil recovery, there might be a reason for choosing one rather than another. Note that in graphical analysis, any place where the curve crosses the line y=x is a point that stops the orbit--it is a fixed point of this dynamic geometric system. Each curve in Figure 5.8 has three intersection points, F, F', and F" that are fixed points. Use these to separate the possible seed values on the x-axis into three sets (drop a vertical line from each of F, F', and F" to the x-axis to do so). This partition completely determines the behavior of all seed values relative to the curve.

Figures 5.8 a,b,c,d suggest that orbits are pushed from F' for any seed value on either side of F' (between F and F"). This suggests that, in the case of semidesert soils, human planting is appropriate at a soil thickness immediately to the left of F' to enlarge the size of the interval between F' and F". Such enlargement is important; within this interval, an increase in the x-coordinate soil thickness results in an increase in the y-coordinate production of new soil. Orbits to the right of F' are staircases that exhibit various geometric dynamics as they approach F". Orbits to the left of F' are descending staircases; a decrease in soil thickness results in lowering the production of new soil, because the soil is stripped of its own capacity to build itself. Thus intervention slightly to the left of F' would slide this fixed point to the left, which reduces the unwanted dynamic in the interval to the left of F' and increases the desirable dynamic to the right of F'.

Orbit shapes suggest the extent to which control can be retained over the geometric/soil dynamics of a proposed insertion of plants. The most geometric control appears in Figures 5.8a and 5.8b where the general pattern and direction of feedback are clear. Least control shows in Figure 5.8c, and a return to regular control, in a different form appears in Figure 5.8d. It makes a difference which of the splines is chosen to represent the delicate feedback mechanisms in semidesert soil. If the shade plants chosen had only a quick effect, in which case the curve would rise and then drop sharply after the initial planting (Figure 5.8c), the corresponding geometric dynamics of feedback would fly out of control. The orbit would bounce alternately from one end of a rectangular tunnel to the other. In this case, the lack of geometric control reflects a lack of real-world control--the loss of control of terrestrial feedback in semidesert soil production, caused by the introduction of quick-fix types of plantings.

References

1. Arlinghaus, S., Nystuen, J., and Woldenberg, M. An application of graphical analysis to semidesert soils. *Geographical Review*, July 1992, Vol. 82, No. 3, 244-252.
2. Devaney, R. L., and Keen, L. eds. *Chaos and fractals: The mathematics behind the computer graphics,* Proceedings of symposia in applied mathematics, vol. 39, American Mathematical Society, Providence, RI, 1989.
3. Feigenbaum, M. J. 1980. Universal behavior in non-linear systems. *Los Alamos Science,* summer, 1980, 4-27.
4. Gleick, J. *Chaos: Making a new science,* Penguin Books, New York, 1987.

CHAPTER 6

EDUCATION DATA ANALYSIS

ANALYTICAL TECHNIQUES/TOOLS USED:

Straight line curve-fitting--least squares
Residual plots
Root Mean Square error

DATA TYPE: VARIABLE LONGITUDINAL DATA

transfer of data from data base to spreadsheet
cleaning, analysis, and graphing of transferred data

Overview of Data

In national statistics, education of women is often regarded as one of the single best predictors of infant mortality; a woman with even minimal education learns elements of personal care and nutrition, and medical information during pregnancy and early motherhood, that goes beyond natural instinct. Thus, education of females is a good indicator of infant mortality, particularly in developing countries. Data sets, on infant mortality and adult literacy, female, are exhibited below, derived from the World Resources Institute data base. We use these to analyze the extent to which this theme is present in these data sets.

With any data set (presented in electronic or paper format), it is important first to examine the set for interesting or unusual patterns in the display. These patterns often influence decisions in choosing subsets of data and tools to analyze subsets.

PATTERNS IN DATA--WHAT TO LOOK FOR

1. What is the general organizational scheme of the entire set? Is it arranged alphabetically, numerically, or in some other fashion?

2. Are the real-world entries in the Table (nations, states, counties) expressed as comparable units? For example, county data and national data are generally not comparable.

3.   Are the numerical entries in the Table expressed in comparable units? For example, data in one column might measure percentages while data in another column might measure thousands of dollars--these columns would not be comparable.

4. Are there gaps in the data? If so, what is their significance to the questions you wish to have the data answer?

Table 6.1 contains data derived from the World Resources Institute data base concerning infant mortality and adult literacy, female, for a set of nations in 1970.   In the entire database, there are numerous entries that do not appear in Table 6.1; indeed, Table 6.1 is a "cleaned" version of the data base, tailored to looking at the variables under consideration.   To illustrate how the data sets were cleaned, associate them with the four points above.

1. What is the general organizational scheme of the entire set? Is it arranged alphabetically, numerically, or in some other fashion?

The data base can be sorted by country and by variable, in either order;   no more than 1000 series of data can be exported from the electronic version of the data base to a spreadsheet, in a single action of exporting. For detail on how to export, see Chapter 1.

2. Are the real-world entries in the Table (nations, states, counties) expressed as comparable units? For example, county data and national data are generally not comparable.

There were entries for more countries in the infant mortality data than in the literacy variable; the sets were made to match.

3.   Are the numerical entries in the Table expressed in comparable units?   For example, data in one column might measure percentages while data in another column might measure thousands of dollars--these columns would not be comparable.

The data on infant mortality were presented per thousand; they were converted to percentages in order to be comparable to the variable on adult literacy, female, which is given in percentages.

TABLE 6.1 (source: World Resources Institute)

Table ordered on 1970 female literacy

Infant mortality births per thousand; %			Adult Literacy female; %		Linear fit	Residuals
Country	1970	1990	1970	1990	est.	
Somalia	16.2	13.2	1	14	16.9	−0.7
Yemen Arab Rep	18.6	12	1	26.3	16.9	1.7
Afghanistan	20.3	17.2	2	13.9	16.77	3.53
Chad	17.9	13.2	2	17.9	16.77	1.13
Niger	17.6	13.5	2	16.8	16.77	0.83
Saudi Arabia	14	7.1	2	48.1	16.77	−2.77
Burkina Faso	18.5	13.8	3	8.9	16.64	1.86
Nepal	16.4	12.8	3	13.2	16.64	−0.24
Mali	20.6	16.9	4	23.9	16.51	4.09
Senegal	15.4	8.7	5	25.1	16.38	−0.98
Central Afr Rep	15	10.4	6	24.9	16.25	−1.25
Guinea−Bissau	18.9	15.1	6	24	16.25	2.65
Sudan	15.6	10.8	6	11.7	16.25	−0.65
Angola	18.6	13.7	7	28.5	16.12	2.48
Guinea	22.2	20.8	7	13.4	16.12	6.08
Togo	14.1	9.4	7	30.7	16.12	−2.02
Benin	16	9	8	15.6	15.99	0.01
Liberia	17.3	14.2	8	28.8	15.99	1.31
Sierra Leone	20.4	15.4	8	11.3	15.99	4.41
Yemen, PDR	18.6	12	9	26.1	15.86	2.74
Burundi	14.3	11.9	10	39.8	15.73	−1.43
Cote d'Ivoire	14.3	9.6	10	40.2	15.73	−1.43
Morocco	13.8	8.2	10	38	15.73	−1.93
Algeria	15	7.4	11	45.5	15.6	−0.6
Pakistan	14.5	10.9	11	21.1	15.6	−1.1
Bangladesh	14	11.9	12	22	15.47	−1.47
Libya	13	8.2	13	50.4	15.34	−2.34
Mozambique	17.5	14.1	14	21.3	15.21	2.29
Nigeria	14.6	10.5	14	39.5	15.21	−0.61

Haiti	15	9.7	17	47.4	14.82	0.18
Iran, Islamic Rep	14.5	5.2	17	43.3	14.82	−0.32
Tunisia	13.8	5.2	17	56.3	14.82	−1.02
Ghana	11.7	9	18	51	14.69	−2.99
Iraq	11.1	6.9	18	49.3	14.69	−3.59
Cameroon	13.6	9.4	19	42.6	14.56	−0.96
Congo	11	7.3	19	43.9	14.56	−3.56
Kenya	10.8	7.2	19	58.5	14.56	−3.76
Egypt	17	6.5	20	33.8	14.43	2.57
India	14.5	9.9	20	33.7	14.43	0.07
Syrian Arab Rep	10.7	4.8	20	50.8	14.43	−3.73
Rwanda	14	12.2	21	37.1	14.3	−0.3
Gabon	14.7	10.3	22	48.5	14.17	0.53
Zaire	13.3	8.3	22	60.7	14.17	−0.87
Cambodia	13	13	23	22.4	14.04	−1.04
Papua N Guinea	13	5.9	24	37.8	13.91	−0.91
Jordan	10.2	4.4	29	70.3	13.26	−3.06
Uganda	11.8	10.3	30	34.9	13.13	−1.33
Turkey	15.3	7.6	34	71	12.61	2.69
Guatemala	10.8	5.9	37	47.1	12.22	−1.42
Zambia	11.5	8	37	65.3	12.22	−0.72
Indonesia	12.4	7.5	42	68	11.57	0.83
Kuwait	5.5	1.8	42	66.7	11.57	−6.07
Madagascar	19.5	12	43	72.9	11.44	8.06
Botswana	11	6.7	44	65.1	11.31	−0.31
Bolivia	15.7	11	46	70.7	11.05	4.65
Zimbabwe	10.1	6.6	47	60.3	10.92	−0.82
Malaysia	5	2.4	48	70.4	10.79	−5.79
Honduras	12.3	6.9	50	70.6	10.53	1.77
El Salvador	11.7	6.4	53	70	10.14	1.56
Myanmar	11	7	57	72.3	9.62	1.38
Lebanon	5.2	4.8	58	73.1	9.49	−4.29
Peru	12.6	8.8	60	78.7	9.23	3.37
Brazil	10	6.3	63	79.8	8.84	1.16

Dominican Rep	10.5	6.5	65	81.8	8.58	1.92
Portugal	6.1	1.5	65	81.5	8.58	−2.48
Ecuador	10.7	6.3	68	83.8	8.19	2.51
Mexico	7.9	4.3	69	85.1	8.06	−0.16
Sri Lanka	6.1	2.8	69	83.5	8.06	−1.96
Venezuela	6	3.6	71	89.6	7.8	−1.8
Thailand	8.4	2.8	72	89.9	7.67	0.73
Paraguay	6.7	4.2	75	88.1	7.28	−0.58
Colombia	8.2	4	76	85.9	7.15	1.05
Greece	4.2	1.7	76	89.1	7.15	−2.95
Yugoslavia	6.1	2.5	76	88.1	7.15	−1.05
Korea, Rep	5.8	2.5	81	93.5	6.5	−0.7
Panama	5.2	2.3	81	88.2	6.5	−1.3
Philippines	7	4.5	81	89.5	6.5	0.5
Costa Rica	6.6	1.8	87	93.1	5.72	0.88
Cuba	4.9	1.5	87	93	5.72	−0.82
Spain	3.3	1	87	93.4	5.72	−2.42
Chile	9.5	2	88	93.2	5.59	3.91
Guyana	8.2	5.6	89	95.4	5.46	2.74
Argentina	5.6	3.2	92	95.1	5.07	0.53
Italy	3.3	1.1	93	96.4	4.94	−1.64
Uruguay	4.7	2.4	93	95.9	4.94	−0.24
Jamaica	4.5	1.7	97	98.6	4.42	0.08

Regression Output:

Constant	17.03
Std Err of Y Est	2.479
R Squared	0.727
No. of Observations	86
Degrees of Freedom	84
X Coefficient(s)	−0.13
Std Err of Coef.	0.008

Linear equation: $y = -0.13 * x + 17.03$

4. Are there gaps in the data. If so, what is their significance to the questions you wish to have the data answer?

There were gaps; adult literacy, female, data was only available for the years 1970 and 1990. Infant mortality data sets, both actual and projected, were available from 1955 to 2025, at five year intervals, and from 1970 to 1990 on an annual basis. The two data sets were made to match exactly; using the variable dealing with adult female literacy therefore greatly reduced the size of the infant mortality table--from over 25 columns to 2 columns.

Straight line curve fitting--least squares analysis--1970 data

The pattern of general decline of mortality, as literacy improves, suggests that a straight line might fit the actual data fairly well. In Table 6.1, a straight line has been fit to the actual data from 1970, with percent adult female literacy as the independent variable and percent infant mortality as the dependent variable, using least squares "regression" analysis. The fit is fair, as is indicated by the r-squared value of 0.727 (Table 6.1).

The procedure for obtaining this line is straightforward and uses only the capability of a good spreadsheet. First, sort the entire data range, columns A (country), B (1970 infant mortality data), C (1990 infant mortality data), D (1970 Adult Literacy, female), and E (1990 Adult Literacy, female), using the female literacy, 1970, column as the primary variable sorted in ascending order (with country name as the secondary variable to break ties in the primary variable).

Then, select the regression feature from the spreadsheet. Enter the column, Adult Literacy, 1970, as the x-axis value; enter the infant mortality data for 1970 as the next variable (the Y-range in Lotus 1-2-3, release 2.3). Choose a blank location of the spreadsheet as the output range. Then, go with the regression, and the output should be similar to that part of Table 6.1 below the data entries. Typically, the last two lines, giving the phrase "linear fit" and the equation of the regression line are not part of the output. The user needs to understand the content of the output enough to know that generally, the equation of the line of least squares will be of the form

$$y=mx+b$$

where m is the slope of the line and b is its intercept on the y-axis. The "X Coefficient(s)" in Table 6.1, -0.13, is the x coefficient--m--in the displayed equation above. The "constant" in Table 6.1, 17.03, is the y-intercept and so is the b-value in the displayed equation.

The reader unfamiliar with using a spreadsheet might wish to actually try fitting a straight line by following the instructions below, step by step.

STRAIGHT LINE FIT TO THE DATA OF TABLE 6.1
(Refer to Table 6.1 and Figure 6.1)

1. Enter the values for the independent variable, 1970 female literacy data, in column D, with the first in cell D5.

2. Enter the dependent variable, 1970 infant mortality in column B, with the first entry in cell B5.

3. Choose the regression feature from the software, with the x values as in step 1 and the y values as in step 2.

4. Choose the output range as a blank area in the spreadsheet. Then proceed with the calculation as directed by the software; the output from the regression will appear in a form similar to the one in Table 6.1, bottom half (produced in Lotus 1-2-3, release. 2.3).

5. The equation below the output range must generally be derived by the user from the regression output. The slope-intercept form for the equation of a straight line (y=mx+b) is used. The "X Coefficient" from the regression output is used as "m". The "Constant" from the regression output is used as b.

6. In a separate column of the spreadsheet, labelled "Linear fit est." in Table 6.1 (Column F), enter the equation derived from the regression: in cell F5, enter the formula -0.13*A5+17.03. The value 16.9 should appear at the top of the "Linear fit est." column. Then, copy the cell content from E5 to the cells below it; this should produce the numerical range shown in the "Linear fit est." column.

7. Graph the results; select an XY-graph. Put the entries for 1970 female literacy in the X-range, and the 1970 infant mortality values in the A range. Enter the entries of the estimated linear data in an additional range, B, of the graph. Then "view" the graph; it should appear generally as in Figure 1.1.

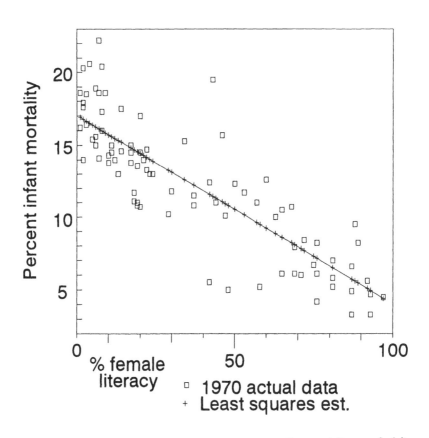

Figure 6.1. Scatter diagram from Table 6.1--data from 1970; percent literacy of adult females as independent variable and percent infant mortality as dependent variable. Least squares line is fit to the scatter.

As the R-squared value suggests, the fit of the line to the scatter of dots is not that tight, although the line does describe the general trend (with various exceptions) that greater literacy, on the whole, is associated with lower infant mortality. It may be that a straight line fit is not the one that best describes this situation, although certainly there appeared to be no other evident choices. When the residuals are plotted (Figure 6.2) there is no evident curvilinear pattern to suggest some non-linear choice. Generally, a residual is the y-distance between the actual and estimated y-components of each x-value. This distance is one with a sign attached as y(estimated) - y(actual) = residual. The entries in Column G, residuals, of Table 6.1 were calculated in this manner. The x-axis is the line of least squares for the residual plot.

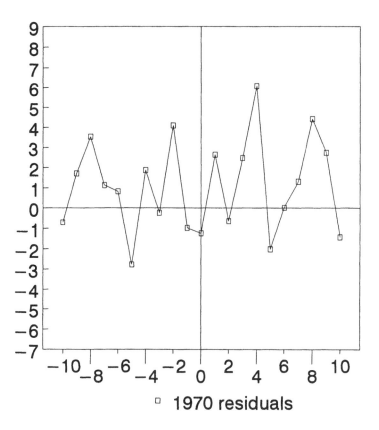

□ 1970 residuals

Figure 6.2. Residual plot, 1970 regression analysis from Table 6.1.

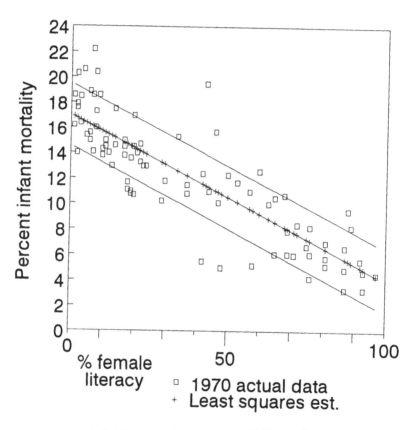

Figure 6.3. Band of width one root mean square error, 1970 regression.

On the basis of the evidence to date, there appears only to be a loose relationship between the selected variables. To see the extent of error in the y-variable, it is useful to use the statistic calculated in the regression, labelled "Std Err of Y Est." In the case of regression, this value is the Root Mean Square Error (RMS). The RMS given in Table 6.1 is 2.479; when this quantity is added and subtracted from the constant of the regression equation, two new equations (with the same slope as the regression line) emerge. Because all three lines have the same slope, they are all parallel. The plotted equations create a band around the regression line which, in a linear fit, contains about 68% of the points in the scatter. For a given value of the independent variable, one would expect that the associated y-value would lie within this band 68% of the time. That appears to be the case with the actual values here, although the band that the RMS takes to capture the 68% is one that is fairly wide (Figure 6.3). The two equations associated with the edges of this band are, on the top:

$$y=-0.13*x+19.509$$

and on the bottom:

$$y=-0.13*x+14.551.$$

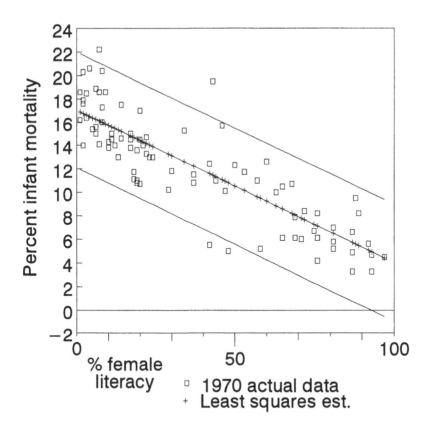

Figure 6.4. Band of width twice the root mean square error, 1970 regression.

When twice the RMS is added and subtracted to the regression constant, another band that tracks the regression line emerges. In a linear fit to a scatter of dots, this band captures about 95% of the scatter, as it does in this case (Figure 6.4). The two equations associated with the edges of this band are, on the top:

$$y=-0.13*x+21.988$$

and on the bottom:

$$y=-0.13*x+12.072.$$

One might wonder which countries are the outliers, beyond the two root mean square error band, and check back in the tabular material to find out; this sort of analysis can highlight countries that do not seem to fit the general trend.

A logical next question is to consider whether or not the fit gets better over time--as the education of women accelerates (if it does) is the association stronger? To analyze this situation, repeat for the 1990 data the steps that were carried out on the 1970 data.

Straight line curve fitting--least squares analysis--1990 data

The procedure for obtaining this line is straightforward and uses only the capability of a good spreadsheet. First, sort the entire data range, columns A (country), B (1970 infant mortality data), C (1990 infant mortality data), D (1970 Adult Literacy, female), and E (1990 Adult Literacy, female), using the female literacy, 1990, column as the primary variable sorted in ascending order (with country name as the secondary variable to break ties in the primary variable).

Then, select the regression feature from the spreadsheet. Enter the column, Adult Literacy, 1990, as the x-axis value; enter the infant mortality data for 1990 as the next variable (the Y-range in Lotus 1-2-3, release 2.3). Choose a blank location of the spreadsheet as the output range. Then, go with the regression, and the output should be similar to that part of Table 6.2 below the data entries. Typically, the last two lines, giving the phrase "linear fit" and the equation of the regression line are not part of the output. The user needs to understand the content of the output enough to know that generally, the equation of the line of least squares will be of the form

$$y=mx+b$$

where m is the slope of the line and b is its intercept on the y-axis. The "X Coefficient(s)" in Table 6.2, -0.13, is the x coefficient--m--in the displayed equation above. The "constant" in Table 6.2, 15.23, is the y-intercept and so is the b-value in the displayed equation.

TABLE 6.2 (source: World Resources Institute)

Table ordered on 1990 female literacy

Country	Infant mortality births per thousand; % 1970	1990	Adult Literacy female; % 1970	1990	Linear fit est.	Residuals
Burkina Faso	18.5	13.8	3	8.9	14.07	−0.273
Sierra Leone	20.4	15.4	8	11.3	13.76	1.639
Sudan	15.6	10.8	6	11.7	13.70	−2.909
Nepal	16.4	12.8	3	13.2	13.51	−0.714
Guinea	22.2	20.8	7	13.4	13.48	7.312
Afghanistan	20.3	17.2	2	13.9	13.42	3.777
Somalia	16.2	13.2	1	14	13.41	−0.21
Benin	16	9	8	15.6	13.20	−4.202
Niger	17.6	13.5	2	16.8	13.04	0.454
Chad	17.9	13.2	2	17.9	12.90	0.297
Pakistan	14.5	10.9	11	21.1	12.48	−1.587
Mozambique	17.5	14.1	14	21.3	12.46	1.639
Bangladesh	14	11.9	12	22	12.37	−0.47
Cambodia	13	13	23	22.4	12.31	0.682
Mali	20.6	16.9	4	23.9	12.12	4.777
Guinea−Bissau	18.9	15.1	6	24	12.11	2.99
Central Afr Rep	15	10.4	6	24.9	11.99	−1.593
Senegal	15.4	8.7	5	25.1	11.96	−3.267
Yemen, PDR	18.6	12	9	26.1	11.83	0.163
Yemen Arab Rep	18.6	12	1	26.3	11.81	0.189
Angola	18.6	13.7	7	28.5	11.52	2.175
Liberia	17.3	14.2	8	28.8	11.48	2.714
Togo	14.1	9.4	7	30.7	11.23	−1.839
India	14.5	9.9	20	33.7	10.84	−0.949
Egypt	17	6.5	20	33.8	10.83	−4.336
Uganda	11.8	10.3	30	34.9	10.69	−0.393
Rwanda	14	12.2	21	37.1	10.40	1.793
Papua N Guinea	13	5.9	24	37.8	10.31	−4.416
Morocco	13.8	8.2	10	38	10.29	−2.09

Nigeria	14.6	10.5	14	39.5	10.09	0.405
Burundi	14.3	11.9	10	39.8	10.05	1.844
Cote d'Ivoire	14.3	9.6	10	40.2	10.00	−0.404
Cameroon	13.6	9.4	19	42.6	9.692	−0.292
Iran, Islamic Rep	14.5	5.2	17	43.3	9.601	−4.401
Congo	11	7.3	19	43.9	9.523	−2.223
Algeria	15	7.4	11	45.5	9.315	−1.915
Guatemala	10.8	5.9	37	47.1	9.107	−3.207
Haiti	15	9.7	17	47.4	9.068	0.632
Saudi Arabia	14	7.1	2	48.1	8.977	−1.877
Gabon	14.7	10.3	22	48.5	8.925	1.375
Iraq	11.1	6.9	18	49.3	8.821	−1.921
Libya	13	8.2	13	50.4	8.678	−0.478
Syrian Arab Rep	10.7	4.8	20	50.8	8.626	−3.826
Ghana	11.7	9	18	51	8.6	0.4
Tunisia	13.8	5.2	17	56.3	7.911	−2.711
Kenya	10.8	7.2	19	58.5	7.625	−0.425
Zimbabwe	10.1	6.6	47	60.3	7.391	−0.791
Zaire	13.3	8.3	22	60.7	7.339	0.961
Botswana	11	6.7	44	65.1	6.767	−0.067
Zambia	11.5	8	37	65.3	6.741	1.259
Kuwait	5.5	1.8	42	66.7	6.559	−4.759
Indonesia	12.4	7.5	42	68	6.39	1.11
El Salvador	11.7	6.4	53	70	6.13	0.27
Jordan	10.2	4.4	29	70.3	6.091	−1.691
Malaysia	5	2.4	48	70.4	6.078	−3.678
Honduras	12.3	6.9	50	70.6	6.052	0.848
Bolivia	15.7	11	46	70.7	6.039	4.961
Turkey	15.3	7.6	34	71	6	1.6
Myanmar	11	7	57	72.3	5.831	1.169
Madagascar	19.5	12	43	72.9	5.753	6.247
Lebanon	5.2	4.8	58	73.1	5.727	−0.927
Peru	12.6	8.8	60	78.7	4.999	3.801
Brazil	10	6.3	63	79.8	4.856	1.444

Portugal	6.1	1.5	65	81.5	4.635	−3.135
Dominican Rep	10.5	6.5	65	81.8	4.596	1.904
Sri Lanka	6.1	2.8	69	83.5	4.375	−1.575
Ecuador	10.7	6.3	68	83.8	4.336	1.964
Mexico	7.9	4.3	69	85.1	4.167	0.133
Colombia	8.2	4	76	85.9	4.063	−0.063
Paraguay	6.7	4.2	75	88.1	3.777	0.423
Yugoslavia	6.1	2.5	76	88.1	3.777	−1.277
Panama	5.2	2.3	81	88.2	3.764	−1.464
Greece	4.2	1.7	76	89.1	3.647	−1.947
Philippines	7	4.5	81	89.5	3.595	0.905
Venezuela	6	3.6	71	89.6	3.582	0.018
Thailand	8.4	2.8	72	89.9	3.543	−0.743
Cuba	4.9	1.5	87	93	3.14	−1.64
Costa Rica	6.6	1.8	87	93.1	3.127	−1.327
Chile	9.5	2	88	93.2	3.114	−1.114
Spain	3.3	1	87	93.4	3.088	−2.088
Korea, Rep	5.8	2.5	81	93.5	3.075	−0.575
Argentina	5.6	3.2	92	95.1	2.867	0.333
Guyana	8.2	5.6	89	95.4	2.828	2.772
Uruguay	4.7	2.4	93	95.9	2.763	−0.363
Italy	3.3	1.1	93	96.4	2.698	−1.598
Jamaica	4.5	1.7	97	98.6	2.412	−0.712

Regression Output:

Constant	15.23
Std Err of Y Est	2.340
R Squared	0.718
No. of Observations	86
Degrees of Freedom	84
X Coefficient(s)	−0.13
Std Err of Coef.	0.009

Linear equation: $y = -0.13*x + 15.23$

The reader unfamiliar with using a spreadsheet might wish to actually try fitting a straight line by following the instructions below, step by step.

STRAIGHT LINE FIT TO THE DATA OF TABLE 6.2
(Refer to Table 6.2 and Figure 6.5)

1. Enter the values for the independent variable, 1990 female literacy data, in column D, with the first in cell D5.

2. Enter the dependent variable, 1990 infant mortality in column B, with the first entry in cell B5.

3. Choose the regression feature from the software, with the x values as in step 1 and the y values as in step 2.

4. Choose the output range as a blank area in the spreadsheet. Then proceed with the calculation as directed by the software; the output from the regression will appear in a form similar to the one in Table 6.2, bottom half (produced in Lotus 1-2-3, release. 2.3).

5. The equation below the output range must generally be derived by the user from the regression output. The slope-intercept form for the equation of a straight line (y=mx+b) is used. The "X Coefficient" from the regression output is used as "m". The "Constant" from the regression output is used as b.

6. In a separate column of the spreadsheet, labelled "Linear fit est." in Table 6.2 (Column F), enter the equation derived from the regression: in cell F5, enter the formula -0.13*A5+15.23. The value 14.07 should appear at the top of the "Linear fit est." column. Then, copy the cell content from E5 to the cells below it; this should produce the numerical range shown in the "Linear fit est." column.

7. Graph the results; select an XY-graph. Put the entries for 1990 female literacy in the X-range, and the 1990 infant mortality values in the A range. Enter the entries of the estimated linear data in an additional range, B, of the graph. Then "view" the graph; it should appear generally as in Figure 6.5.

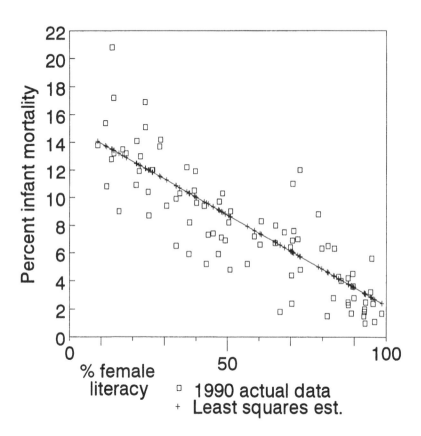

Figure 6.5. Scatter diagram from Table 6.2--data from 1990; percent literacy of adult females as independent variable and percent infant mortality as dependent variable. Least squares line is fit to the scatter.

As the R-squared value suggests, the fit of the line to the scatter of dots is not that tight, although the line does describe the general trend (with various exceptions) that greater literacy, on the whole, is associated with lower infant mortality. Indeed, the R-squared suggests that the 1990 regression line is no better a fit to the underlying scatter than was the 1970 regression line.

When the residuals are plotted (Figure 6.6) there is no evident curvilinear pattern to suggest some non-linear choice. Generally, a residual is the y-distance between the actual and estimated y-components of each x-value. This distance is one with a sign attached as y(estimated) - y(actual) = residual. The entries in Column G, residuals, of Table 6.2 were calculated in this manner. The x-axis is the line of least squares for the residual plot.

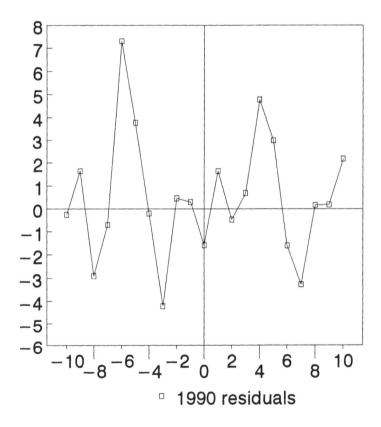

□ 1990 residuals

Figure 6.6. Residual plot, 1990 regression analysis from Table 6.2.

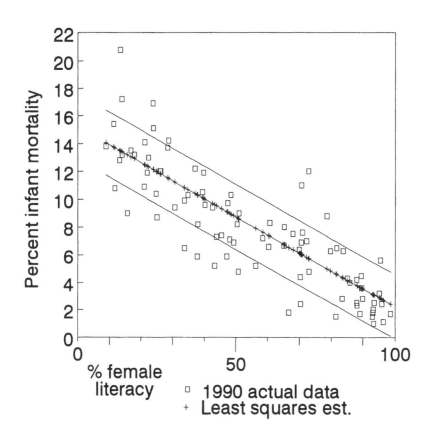

Figure 6.7. Band of width one root mean square error, 1990 regression.

On the basis of the evidence to date, there appears only to be a loose relationship between the selected variables. To see the extent of error in the y-variable, it is useful to use the statistic calculated in the regression, labelled "Std Err of Y Est." In the case of regression, this value is the Root Mean Square Error (RMS). The RMS given in Table 6.2 is 2.340; when this quantity is added and subtracted from the constant of the regression equation, two new equations (with the same slope as the regression line) emerge. Because all three lines have the same slope, they are all parallel. The plotted equations create a band around the regression line which, in a linear fit, contains about 68% of the points in the scatter. One would expect, given a value of the independent variable, that the associated y-value would lie within this band about 68% of the time. That appears to be the case with the actual values here, although the band that the RMS takes to capture the 68% is fairly wide (Figure 6.7). The two equations associated with the edges of this band are, on the top:

$$y = -0.13*x + 17.57$$

and on the bottom:

$$y = -0.13*x + 12.89.$$

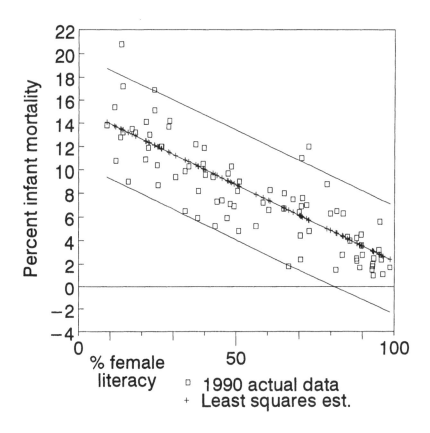

Figure 6.8. Band of width twice the root mean square error, 1990 regression.

When twice the RMS is added and subtracted to the regression constant, another band that tracks the regression line emerges. In a linear fit to a scatter of dots, this band captures about 95% of the scatter, as it does in this case (Figure 6.8). The two equations associated with the edges of this band are, on the top:

$$y = -0.13 * x + 19.91$$

and on the bottom:

$$y = -0.13 * x + 10.55.$$

The line fit to this scatter seems a reasonable choice; the scatter of dots is readily partitioned using the RMS--a value that is to the regression line in a linear fit as the standard deviation is to the mean in a normal distribution. The 1990 RMS makes a slightly tighter band around the regression line than does the 1970 RMS.

BLACK BOX SUMMARY
see Introduction for theoretical explanation

LEAST SQUARES REGRESSION LINE

$$y = mx + b$$

where
m is the slope of the line, or the "x-coefficient"
b is the y-intercept, or the "constant."

A view of the two least squares fit
    Field evidence over a number of years and in various locales points to the strong relationship between the education of women and infant mortality. There is little difference between the 1970 and 1990 graphs-- the 1990 graph reflects the general increase in the educational level of women; both have the same slope. Indeed, even when only African countries and only South American countries were analyzed in this manner, similar situations arose. Basically, the data show the general trend.
    If allocation of reserves is to be made based on arguments supported by graphs such as these, then these graphs must be accompanied by strong supportive explanations. There may well be some important reasons that the graphs do not support the field evidence in this case as strongly as one might suspect.

1. There is very little data from developed countries in Tables 6.1 and 6.2; most of it is from developing countries. Therefore, numerous data points are omitted which would have high literacy and low mortality coordinates. Had these been included, they would reduce the band width of the root mean square and tighten the fit. The reason they were not included was that there was incomplete data for them. Thus, it is to the advantage of researchers wishing to make arguments about the importance of education for women to encourage agencies to obtain data from developed, as well as developing, countries.

2. Literacy is used as a variable to measure "education;" this variable does not account for certain sorts of education, such as training in various matters that relate to public health--as might be introduced in developing countries by various international agencies, that may be acquired without being literate. Education of this sort also reduces infant mortality rates, yet it would not be counted within the variable. Thus, there are instances in which the literacy rate is low while the infant mortality rate is surprisingly low. Data points of this sort contribute to the spread of the root mean square band. Researchers might wish to elaborate on the importance of having longitudinal data cutting across various skills and training levels as measures of education.

3. Finally, there is a host of other factors associated with infant mortality; one might argue that to have any single factor (possibly confounded with others), such as literacy of women, show an R-squared value of over 0.70, is indeed good support for the field observations.

CHAPTER 7

TRANSPORTATION AND COMMUNICATION
DATA ANALYSIS

ANALYTICAL TECHNIQUES/TOOLS USED:

Historical maps
Space-filling measured by density
Rank-ordering
Fitting maps to empirical curves

DATA TYPE: ABUNDANT

historical evidence

Overview of Data

One vision of the world at the beginning of the third millennium, might include a planet working to build its infrastructure within existing national boundaries and indeed, searching for political solutions to the difficulties in extending elements of the infrastructure across national boundaries. Computerization can free human minds from the mass of detail involved in monitoring and planning such systems, so that they might focus on the deeper abstract philosophical, ethical, and political issues involved in decision making. Within such a world view, the updating of various enduring ideas, within the current technological environment, would be instrumental in freeing human time at managerial levels, and in creating smoothly functioning, accurately planned, spatial layouts for infrastructures and various networks. Jobs at the planning level might become even more competitive, with the better positions going to individuals conversant with, or well-trained in, the principles associated with these issues. Indeed, at a different scale, even the model train community, which currently uses computer software to plan complicated layouts, might benefit from an understanding of how various components of the electronic environment (including Geographic Information Systems--GIS) are becoming integrated into the network of full-sized trains.

The railroad industry in the United States is becoming significant both as a user of GIS and as a communications component in the transmission of digital information. Fiber optic cable needs to be laid to

169

link long distances and in such a manner that direct physical access to it is relatively easy. Using the channels of land devoted to railroad track is one way to gain such access. Further, since 1983, Amtrak (in the United States) has begun using electronic tools (GIS) in a real estate capacity to manage the leasing of its land and air rights and in an engineering capacity to manage facilities information such as track layout, various mechanical and electrical systems, and planning. GIS appears to offer substantial promise for innovative management of railway facilities and consequent efficiency of railnet development in the future. The marriage of transportation and communications networks, in railroad channels, is an exciting prospect; when cast in an electronic environment it seems a fresh and powerful notion. The basic idea, however, is not one that is new.

Historical evidence is important to check to see that projects under investigation are not simply repetitions of what has already been done; even more though, it is a sometimes overlooked source that is rich in ideas. What was once a good idea is often still a good idea; an earlier idea might be brought up-to-date, or cast in a different environmental or technological setting. Studies that engage is this sort of research design often communicate a great deal of information about the world over a long stretch of time.

In this context, we look to the past and to Mark Jefferson's use of buffers surrounding railroad tracks as a way to display, with great visual impact, the density of the railroad network within the various continents of the world (Figures 7.1 to 7.6). Each buffer is 16 kilometers wide (10 miles); Jefferson's paper on "The Civilizing Rails" appeared in 1928. In it, he captured the progress of the railroad building era and the pattern of steam locomotive transport, from 1830 to 1930. Jefferson's map of North America shows a rail pattern in which the eastern half of the USA is extensively covered by rail buffers; the pattern tapers off, in density, as one moves away from the east (Figure 7.1). A similar network dispersion appears in Western Europe. The entire subcontinent is well served; indeed, one sees that most of France, Switzerland, the United Kingdom, Germany, the Low Countries, Denmark, Czechoslovakia, and Hungary lie within 16 kilometers of more than one rail line. The intersection of buffers must be quite extensive to produce the solid white pattern present on the map (Figure 7.2). The remaining maps show rail buffers for Asia (Figure 7.3), South America (Figure 7.4), Africa (Figure 7.5), and Australia (Figure 7.6). Country boundaries are generally not evident; these maps show the broad continental sweep of coverage offered by rail lines.

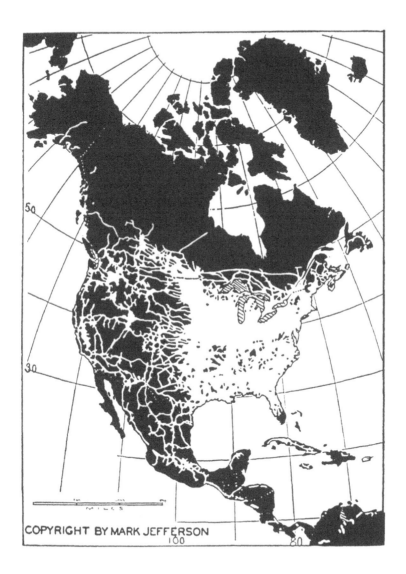

Figure 7.1. Jefferson's view of North America; white buffers are 16 kilometers (10 miles) wide. Reprinted with permission of *Economic Geography*.

In the years following 1928, others made use of ideas similar to those of Jefferson; indeed, Berry and Ginsburg, 1961, created an "Atlas of Economic Development" in which rail density was one indicator of development. Kansky, 1963, used tools from the mathematical subfield of graph theory to characterize the notion of rail density (one of the earliest uses of graph theory in geography). Today, buffers of the sort used by Jefferson to surround rail lines, are common in the electronic environment of the Geographic Information System (GIS) and are critical in looking at the spread in space of styles of land use. The spatial analysis that must have taken Jefferson many days to complete, can now be executed in an electronic environment, and stored there for later modification, in a matter of hours.

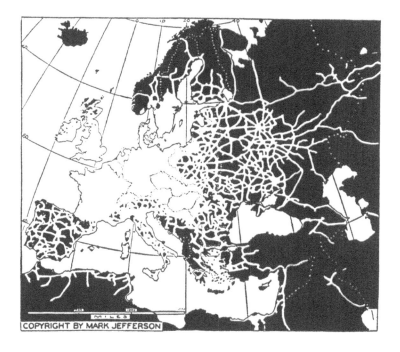

Figure 7.2. Jefferson's view of Europe; white buffers are 16 kilometers (10 miles) wide. Reprinted with permission of *Economic Geography*.

Cleaning of data

In what follows, we take a closer look at current data that is the sort that must have been the sort used by Jefferson. Indeed, data is available in one electronic data base only for the year 1989 The current electronic data will be used in conjunction with the historical visual display; there is more than one way to obtain information on missing entries. With any data set (presented in electronic or paper format), it is important first to examine the set for interesting or unusual patterns in the display. These patterns often influence decisions in choosing subsets of data and tools to analyze subsets.

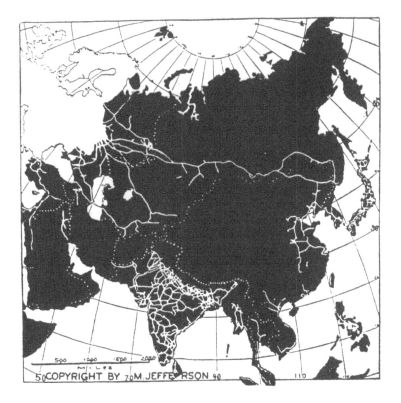

Figure 7.3. Jefferson's view of Asia; buffers are 16 kilometers (10 miles) wide. Reprinted with permission of *Economic Geography*.

Figure 7.4. Jefferson's view of South America; buffers are 16 kilometers (10 miles) wide. Reprinted with permission of *Economic Geography*.

PATTERNS IN DATA--WHAT TO LOOK FOR

1. What is the general organizational scheme of the entire set? Is it arranged alphabetically, numerically, or in some other fashion?

2. Are the real-world entries in the Table (nations, states, counties) expressed as comparable units? For example, county data and national data are generally not comparable.

3. Are the numerical entries in the Table expressed in comparable units? For example, data in one column might measure percentages while data in another column might measure thousands of dollars--these columns would not be comparable.

4. Are there gaps in the data? If so, what is their significance to the questions you wish to have the data answer?

The data base of the World Resources Institute contains data on total extent of track, by country and by continent, for the world in 1989. It also contains data on country area, so that track density, track per unit area, can be calculated. To clean the data in that data base, consider the four points above.

1. What is the general organizational scheme of the entire set? Is it arranged alphabetically, numerically, or in some other fashion?

The World Resources Institute data base can be sorted electronically by variables or by countries; there are more than 500 variables and well over 100 countries for many of the variables. For some types of data, there are extensive time series, too.

2. Are the real-world entries in the Table (nations, states, counties) expressed as comparable units? For example, county data and national data are generally not comparable.

The data for total track are given by country and by continent; thus, one can be used to cut across the other--at different geographic scales.

3.   Are the numerical entries in the Table expressed in comparable units? For example, data in one column might measure percentages while data in another column might measure thousands of dollars--these columns would not be comparable.

The data on total track extent and on land area are both expressed in kilometers; track in kilometers, and land area in kilometers squared. So, the columns are comparable, as they are entered in the data base; no further manipulation is required.

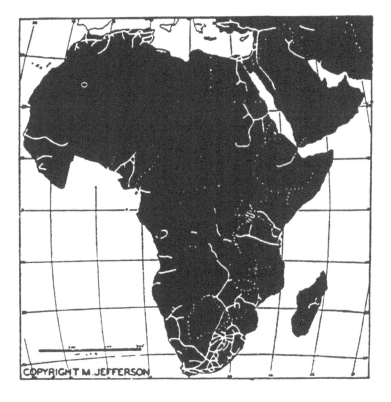

Figure 7.5.   Jefferson's view of Africa; buffers are 16 kilometers (10 miles) wide. Reprinted with permission of heirs of *Economic Geography*.

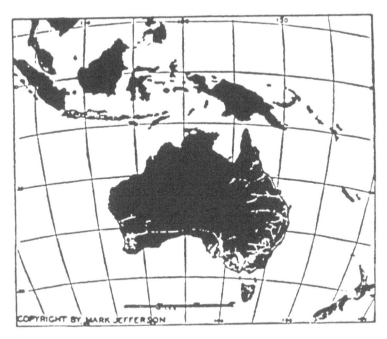

Figure 7.6. Jefferson's view of Oceania; buffers are 16 kilometers (10 miles) wide. Reprinted with permission of *Economic Geography*.

4. Are there gaps in the data? If so, what is their significance to the questions you wish to have the data answer?

There are two gaps; two countries report a total track extent of 0 kilometers. These were discarded from the data base; the distinction between countries reporting "no track" and those that do not have track and do not report it, is not relevant to the calculation of track density. It may be relevant for other studies: such as compliance in reporting data to international data bases.

Calculation of Jefferson buffers

Jefferson surrounded the total extent of railroad track, earlier than 1930, with buffers 10 miles wide (a strip of land 10 miles wide with track running medially down the middle of the strip). Clearly, one could simply use the total track length to calculate track density; it would be a good approximation to the track length times track width. Railroad track gauge varies throughout the world but is generally less than five feet wide (standard gauge in the U.S.A. is 4 feet 8.5 inches; in Russia it is 5 feet). Over a long distance, the difference between the total length, and the total length times five feet, is small.

TABLE 7.1--Jefferson buffers (source: World Res. Ins.)

Alphabetical order by country

	Total track 1989 Km	Track area 1989 Km. sq.	Jefferson buffers	Land area 1989 Km. sq.	Percent Jeff/Area Times 100 DENSITY	Percent Track/Area Times 100 DENSITY
Albania	543	543.814	8688	28750	30.21913	1.891528
Algeria	3787	3792.68	60592	2381740	2.544022	0.159239
Angola	2952	2956.42	47232	1246700	3.788561	0.237140
Argentina	34172	34223.2	546752	2766890	19.76052	1.236885
Australia	39772.4	39832.0	636358.4	7686850	8.278532	0.518184
Austria	5747.5	5756.12	91960	83850	109.6720	6.864783
Bangladesh	2745	2749.11	43920	144000	30.5	1.909109
Belgium	3568	3573.35	57088	33100	172.4712	10.79562
Benin	578	578.867	9248	112620	8.211685	0.514000
Bolivia	3652	3657.47	58432	1098580	5.318866	0.332927
Botswana	713	714.069	11408	581730	1.961047	0.122749
Brazil	30038	30083.0	480608	8511970	5.646260	0.353420
Bulgaria	4300	4306.45	68800	110910	62.03227	3.882832
Burkina Faso	495	495.742	7920	274200	2.888402	0.180795
Cambodia	649	649.973	10384	181040	5.735749	0.359022
Cameroon	1104	1105.65	17664	475440	3.715295	0.232554
Canada	83477.7	83602.9	1335644.	9976140	13.38838	0.838029
Chile	7616	7627.42	121856	756950	16.09828	1.007652
China	54000	54081	864000	9596960	9.002850	0.563522
Colombia	2761	2765.14	44176	1138910	3.878796	0.242788
Congo	517	517.775	8272	342000	2.418713	0.151396
Costa Rica	700	701.05	11200	51100	21.91780	1.371917
Cote d'Ivoire	660	660.99	10560	322460	3.274824	0.204983
Cuba	12795	12814.1	204720	110860	184.6653	11.55889
Czechoslovakia	13103	13122.6	209648	127870	163.9540	10.26249
Denmark	2025	2028.03	32400	43070	75.22637	4.708700
Djibouti	100	100.15	1600	23200	6.896551	0.431681
Dominican Rep	517	517.775	8272	48730	16.97516	1.062539
Ecuador	965.5	966.948	15448	283560	5.447876	0.341003
Egypt	4548	4554.82	72768	1001450	7.266263	0.454822
El Salvador	601.6	602.502	9625.6	21040	45.74904	2.863604
Ethiopia	681	682.021	10896	1221900	0.891726	0.055816
Fiji	644	644.966	10304	18270	56.39846	3.530191
Finland	5863	5871.79	93808	338130	27.74317	1.736549
France	34431	34482.6	550896	551500	99.89048	6.252519
Gabon	668	669.002	10688	267670	3.992976	0.249935
German Dem Re	14007	14028.0	224112	108330	206.8789	12.94933
Germany, Fed R	29206.8	29250.6	467308.8	248580	187.9913	11.76708
Ghana	953	954.429	15248	238540	6.392219	0.400112
Greece	2479	2482.71	39664	131990	30.05076	1.880989
Guatemala	884	885.326	14144	108890	12.98925	0.813046
Guinea	1048	1049.57	16768	245860	6.820141	0.426898

Haiti	40	40.06	640	27750	2.306306	0.144360
Honduras	785	786.177	12560	112090	11.20528	0.701380
Hungary	7703	7714.55	123248	93030	132.4819	8.292544
India	124222.	124408.	1987561.	3287260	60.46256	3.784578
Indonesia	6458	6467.68	103328	1904570	5.425266	0.339587
Iran, Islamic Rep	4667	4674.00	74672	1648000	4.531067	0.283616
Iraq	3081	3085.62	49296	438320	11.24657	0.703965
Ireland	1947	1949.92	31152	70280	44.32555	2.774502
Israel	520	520.78	8320	20770	40.05777	2.507366
Italy	18790.2	18818.3	300643.2	301270	99.79194	6.246352
Jamaica	294	294.441	4704	10990	42.80254	2.679171
Japan	26478.1	26517.8	423649.6	377800	112.1359	7.019009
Jordan	789	790.183	12624	89210	14.15087	0.885756
Kenya	3034	3038.55	48544	580370	8.364319	0.523554
Korea, DPR	8500	8512.75	136000	120540	112.8256	7.062178
Korea, Rep	3149	3153.72	50384	99020	50.88264	3.184935
Lebanon	222	222.333	3552	10400	34.15384	2.137817
Lesotho	1.6	1.6024	25.6	30350	0.084349	0.005279
Liberia	490	490.735	7840	111370	7.039597	0.440634
Madagascar	1054	1055.58	16864	587040	2.872717	0.179814
Malawi	797	798.195	12752	118480	10.76299	0.673696
Malaysia	1672	1674.50	26752	329750	8.112812	0.507811
Mali	641	641.961	10256	1240190	0.826970	0.051763
Mauritania	689	690.033	11024	1025520	1.074966	0.067286
Mexico	20306	20336.4	324896	1958200	16.59156	1.038528
Mongolia	1802	1804.70	28832	1566500	1.840536	0.115206
Morocco	1893	1895.83	30288	446550	6.782667	0.424552
Mozambique	3271	3275.90	52336	801590	6.529023	0.408676
Myanmar	3137	3141.70	50192	676550	7.418816	0.464371
Namibia	2349	2352.52	37584	824290	4.559560	0.285399
Nepal	96	96.144	1536	140800	1.090909	0.068284
Netherlands	2848	2852.27	45568	37330	122.0680	7.640696
New Zealand	4227	4233.34	67632	270990	24.95737	1.562175
Nicaragua	300	300.45	4800	130000	3.692307	0.231115
Nigeria	3505	3510.25	56080	923770	6.070775	0.379992
Norway	4044	4050.06	64704	323900	19.97653	1.250406
Pakistan	8775	8788.16	140400	796100	17.63597	1.103901
Panama	235	235.352	3760	77080	4.878048	0.305335
Paraguay	441	441.661	7056	406750	1.734726	0.108583
Peru	3472.2	3477.40	55555.2	1285220	4.322621	0.270569
Philippines	1058.8	1060.38	16940.8	300000	5.646933	0.353462
Poland	26644	26683.9	426304	312680	136.3387	8.533953
Portugal	3608	3613.41	57728	92390	62.48295	3.911042
Romania	11083	11099.6	177328	237500	74.66442	4.673526
Saudi Arabia	875	876.312	14000	2149690	0.651256	0.040764
Senegal	904	905.356	14464	196720	7.352582	0.460225
Sierra Leone	84	84.126	1344	71740	1.873431	0.117265
Singapore	36	36.054	576	620	92.90322	5.815161
South Africa	23558	23593.3	376928	1221040	30.86942	1.932232
Spain	14438.3	14459.9	231012.8	504780	45.76504	2.864605
Sri Lanka	1453	1455.17	23248	65610	35.43362	2.217923
Sudan	4954	4961.43	79264	2505810	3.163208	0.197997

Swaziland	515	515.772	8240	17360	47.46543	2.971039
Sweden	11660	11677.4	186560	449960	41.46146	2.595228
Switzerland	4497.33	4504.07	71957.28	41290	174.2728	10.90839
Syrian Arab Rep	1771	1773.65	28336	185180	15.30186	0.957801
Tanzania	3569	3574.35	57104	945090	6.042175	0.378202
Thailand	3924	3929.88	62784	513120	12.23573	0.765880
Togo	525	525.787	8400	56790	14.79133	0.925845
Tunisia	1907	1909.86	30512	163610	18.64922	1.167325
Turkey	8430	8442.64	134880	779450	17.30450	1.083154
Uganda	1232	1233.84	19712	235880	8.356791	0.523082
United Kingdom	16894	16919.3	270304	244880	110.3822	6.909237
United States	241778.	242140.	3868449.	9372610	41.27398	2.583493
Uruguay	3002	3006.50	48032	177410	27.07400	1.694663
Venezuela	362.6	363.143	5801.6	912050	0.636105	0.039816
Viet Nam	2900	2904.35	46400	331690	13.98896	0.875621
Yugoslavia	9349	9363.02	149584	255800	58.47693	3.660290
Zaire	5138	5145.70	82208	2345410	3.505058	0.219394
Zambia	2164	2167.24	34624	752610	4.600523	0.287964
Zimbabwe	2759	2763.13	44144	390580	11.30216	0.707444
AFRICA	83837.6	83963.3	1341401.	30306750	4.426081	0.277045
ASIA	271410.	271817.	4342568	28101970	15.45289	0.967254
EUROPE	249051.	249424.	3984818.	4876570	81.71354	5.114757
NORTH & CEN	362713.	363257.	5803415.	22406940	25.90007	1.621183
OCEANIA	44643.4	44710.3	714294.4	8510450	8.393144	0.525358
SOUTH AMER	86482.3	86612.0	1383716.	17818700	7.765531	0.486073
WORLD	1245746	1247614	19931942	1.3E+08	14.82771	0.928122
U.S.S.R.	147608	147829.	2361728	22402200	10.54239	0.659887

Jefferson's choice of a ten mile buffer is reminiscent of earlier U.S. give-away programs in the 1800s designed as an inducement to the railroad industry. Wide swaths of land, including "desirable" land, was given to the railroad industry which was then free to sell the "desirable" land to raise funds to increase track length.

Table 7.1 shows Jefferson buffers for each of the countries in the data base for which there is data on the variables, Total track and Land area. This table is arranged in alphabetical order by country. Column A gives the country name; Column B shows total track, in 1989; Column C shows total track area--total track length multiplied by 1.0015, assuming track width of five feet, or 0.0015 kilometers; Column D shows Jefferson buffers with widths obtained by multiplying the total track length by 16 kilometers; Column E shows land area; Column F shows a measure of the extent to which the Jefferson buffers fill space; and, Column G shows the extent to which the track area fills space--track area divided by land area times 100.

The Jefferson buffers, Column F, offer one way to characterize track density; track density, Column G, offers another. The values in Column G are exactly 16 times those in Column F--this is not surprising, since the width of the buffers is 16 kilometers. Indeed, there is perfect correlation between track density and Jefferson density. Again, this relationship is to be expected since the buffers are dependent on track length; it is noted here simply to reinforce the fact that the buffers are indeed a measure of track density, in addition to being a clear graphic display. Thus, from the standpoint of correlation, it is irrelevant whether one uses Jefferson density or track density. However, there may be good reasons for looking at one density index in some situations and at the other in different situations, even though they are of course mathematically equivalent.

Column F in Table 7.1 is labelled as a "percent" and indeed, it is calculated as a percentage, obtained by dividing the figure for the Jefferson buffers (showing coverage of land area by the buffers) by the actual land area and multiplying by 100. Spain is listed in Table 7.1 as having a value of 45.76504 on this density index; a glance at Figure 7.2 suggests that even in 1928 this value of about 45% might have seemed reasonable--somewhat less than half of Spain appears to be covered by the wiggly white railroad buffers. Belgium and Germany are totally covered by buffers in Figure 7.2, as are various other Western European countries; this means that the buffers must overlap and that many locations within these countries are within 10 miles of more than one railroad line. The entries in Table 7.1 for Belgium of 172.4712, for Germany (partitioned) of 206.8789 and 187.9913, and others, reflect this overlap in buffer coverage. Whenever the percent in Column F exceeds 100, the track density, measured using Jefferson buffers, fills the entire country. The excess over 100 reflects the extent of overlap.

The Jefferson buffers are useful visually; they do not communicate a great deal when they fill all the space. In this latter situation, it may make more sense to look at the density of the actual track (Column G in Table 7.1): for Belgium it is 10.79562 and for Germany (partitioned) it is 12.94933 and 11.76708. What this means is that about 10% of the land of Belgium (and 11 or 12% of the land of Germany) is devoted to railroad tracks.

When track coverage is sparse or moderate, the Jefferson buffers offer a good view of density; when these buffers fill space, the numerical values associated with them is the only way to make distinctions, and indeed, the equivalent characterization of track area may be better yet. What this suggests at a deeper level is that even if one has perfect correlation between two variables, it may be appropriate to use one at one geographic scale and another at another geographic scale; they are equivalent in one sense, but not in another.

TABLE 7.2 (source World Resources Institute)

Countries and continents ordered by Jefferson densities

	Total track 1989 Km	Track area 1989 Km. sq.	Jefferson buffers	Land area 1989 Km. sq.	Percent Jeff/Area Times 100 DENSITY	Percent Track/Area Times 100 DENSITY
German Dem Re	14007	14028.0	224112	108330	206.8789	12.94933
Germany, Fed R	29206.8	29250.6	467308.8	248580	187.9913	11.76708
Cuba	12795	12814.1	204720	110860	184.6653	11.55889
Switzerland	4497.33	4504.07	71957.28	41290	174.2728	10.90839
Belgium	3568	3573.35	57088	33100	172.4712	10.79562
Czechoslovakia	13103	13122.6	209648	127870	163.9540	10.26249
Poland	26644	26683.9	426304	312680	136.3387	8.533953
Hungary	7703	7714.55	123248	93030	132.4819	8.292544
Netherlands	2848	2852.27	45568	37330	122.0680	7.640696
Korea, DPR	8500	8512.75	136000	120540	112.8256	7.062178
Japan	26478.1	26517.8	423649.6	377800	112.1359	7.019009
United Kingdom	16894	16919.3	270304	244880	110.3822	6.909237
Austria	5747.5	5756.12	91960	83850	109.6720	6.864783
France	34431	34482.6	550896	551500	99.89048	6.252519
Italy	18790.2	18818.3	300643.2	301270	99.79194	6.246352
Singapore	36	36.054	576	620	92.90322	5.815161
EUROPE	249051.	249424.	3984818.	4876570	81.71354	5.114757
Denmark	2025	2028.03	32400	43070	75.22637	4.708700
Romania	11083	11099.6	177328	237500	74.66442	4.673526
Portugal	3608	3613.41	57728	92390	62.48295	3.911042
Bulgaria	4300	4306.45	68800	110910	62.03227	3.882832
India	124222.	124408.	1987561.	3287260	60.46256	3.784578
Yugoslavia	9349	9363.02	149584	255800	58.47693	3.660290
Fiji	644	644.966	10304	18270	56.39846	3.530191
Korea, Rep	3149	3153.72	50384	99020	50.88264	3.184935
Swaziland	515	515.772	8240	17360	47.46543	2.971039
Spain	14438.3	14459.9	231012.8	504780	45.76504	2.864605
El Salvador	601.6	602.502	9625.6	21040	45.74904	2.863604
Ireland	1947	1949.92	31152	70280	44.32555	2.774502
Jamaica	294	294.441	4704	10990	42.80254	2.679171
Sweden	11660	11677.4	186560	449960	41.46146	2.595228
United States	241778.	242140.	3868449.	9372610	41.27398	2.583493
Israel	520	520.78	8320	20770	40.05777	2.507366
Sri Lanka	1453	1455.17	23248	65610	35.43362	2.217923
Lebanon	222	222.333	3552	10400	34.15384	2.137817
South Africa	23558	23593.3	376928	1221040	30.86942	1.932232
Bangladesh	2745	2749.11	43920	144000	30.5	1.909109
Albania	543	543.814	8688	28750	30.21913	1.891528
Greece	2479	2482.71	39664	131990	30.05076	1.880989
Finland	5863	5871.79	93808	338130	27.74317	1.736549
Uruguay	3002	3006.50	48032	177410	27.07400	1.694663

NORTH & CEN	362713.	363257.	5803415.	22406940	25.90007	1.621183
New Zealand	4227	4233.34	67632	270990	24.95737	1.562175
Costa Rica	700	701.05	11200	51100	21.91780	1.371917
Norway	4044	4050.06	64704	323900	19.97653	1.250406
Argentina	34172	34223.2	546752	2766890	19.76052	1.236885
Tunisia	1907	1909.86	30512	163610	18.64922	1.167325
Pakistan	8775	8788.16	140400	796100	17.63597	1.103901
Turkey	8430	8442.64	134880	779450	17.30450	1.083154
Dominican Rep	517	517.775	8272	48730	16.97516	1.062539
Mexico	20306	20336.4	324896	1958200	16.59156	1.038528
Chile	7616	7627.42	121856	756950	16.09828	1.007652
ASIA	271410.	271817.	4342568	28101970	15.45289	0.967254
Syrian Arab Rep	1771	1773.65	28336	185180	15.30186	0.957801
WORLD	1245746	1247614	19931942	1.3E+08	14.82771	0.928122
Togo	525	525.787	8400	56790	14.79133	0.925845
Jordan	789	790.183	12624	89210	14.15087	0.885756
Viet Nam	2900	2904.35	46400	331690	13.98896	0.875621
Canada	83477.7	83602.9	1335644.	9976140	13.38838	0.838029
Guatemala	884	885.326	14144	108890	12.98925	0.813046
Thailand	3924	3929.88	62784	513120	12.23573	0.765880
Zimbabwe	2759	2763.13	44144	390580	11.30216	0.707444
Iraq	3081	3085.62	49296	438320	11.24657	0.703965
Honduras	785	786.177	12560	112090	11.20528	0.701380
Malawi	797	798.195	12752	118480	10.76299	0.673696
U.S.S.R.	147608	147829.	2361728	22402200	10.54239	0.659887
China	54000	54081	864000	9596960	9.002850	0.563522
OCEANIA	44643.4	44710.3	714294.4	8510450	8.393144	0.525358
Kenya	3034	3038.55	48544	580370	8.364319	0.523554
Uganda	1232	1233.84	19712	235880	8.356791	0.523082
Australia	39772.4	39832.0	636358.4	7686850	8.278532	0.518184
Benin	578	578.867	9248	112620	8.211685	0.514000
Malaysia	1672	1674.50	26752	329750	8.112812	0.507811
SOUTH AMER	86482.3	86612.0	1383716.	17818700	7.765531	0.486073
Myanmar	3137	3141.70	50192	676550	7.418816	0.464371
Senegal	904	905.356	14464	196720	7.352582	0.460225
Egypt	4548	4554.82	72768	1001450	7.266263	0.454822
Liberia	490	490.735	7840	111370	7.039597	0.440634
Djibouti	100	100.15	1600	23200	6.896551	0.431681
Guinea	1048	1049.57	16768	245860	6.820141	0.426898
Morocco	1893	1895.83	30288	446550	6.782667	0.424552
Mozambique	3271	3275.90	52336	801590	6.529023	0.408676
Ghana	953	954.429	15248	238540	6.392219	0.400112
Nigeria	3505	3510.25	56080	923770	6.070775	0.379992
Tanzania	3569	3574.35	57104	945090	6.042175	0.378202
Cambodia	649	649.973	10384	181040	5.735749	0.359022
Philippines	1058.8	1060.38	16940.8	300000	5.646933	0.353462
Brazil	30038	30083.0	480608	8511970	5.646260	0.353420
Ecuador	965.5	966.948	15448	283560	5.447876	0.341003
Indonesia	6458	6467.68	103328	1904570	5.425266	0.339587
Bolivia	3652	3657.47	58432	1098580	5.318866	0.332927
Panama	235	235.352	3760	77080	4.878048	0.305335
Zambia	2164	2167.24	34624	752610	4.600523	0.287964

Namibia	2349	2352.52	37584	824290	4.559560	0.285399
Iran, Islamic Rep	4667	4674.00	74672	1648000	4.531067	0.283616
AFRICA	83837.6	83963.3	1341401.	30306750	4.426081	0.277045
Peru	3472.2	3477.40	55555.2	1285220	4.322621	0.270569
Gabon	668	669.002	10688	267670	3.992976	0.249935
Colombia	2761	2765.14	44176	1138910	3.878796	0.242788
Angola	2952	2956.42	47232	1246700	3.788561	0.237140
Cameroon	1104	1105.65	17664	475440	3.715295	0.232554
Nicaragua	300	300.45	4800	130000	3.692307	0.231115
Zaire	5138	5145.70	82208	2345410	3.505058	0.219394
Cote d'Ivoire	660	660.99	10560	322460	3.274824	0.204983
Sudan	4954	4961.43	79264	2505810	3.163208	0.197997
Burkina Faso	495	495.742	7920	274200	2.888402	0.180795
Madagascar	1054	1055.58	16864	587040	2.872717	0.179814
Algeria	3787	3792.68	60592	2381740	2.544022	0.159239
Congo	517	517.775	8272	342000	2.418713	0.151396
Haiti	40	40.06	640	27750	2.306306	0.144360
Botswana	713	714.069	11408	581730	1.961047	0.122749
Sierra Leone	84	84.126	1344	71740	1.873431	0.117265
Mongolia	1802	1804.70	28832	1566500	1.840536	0.115206
Paraguay	441	441.661	7056	406750	1.734726	0.108583
Nepal	96	96.144	1536	140800	1.090909	0.068284
Mauritania	689	690.033	11024	1025520	1.074966	0.067286
Ethiopia	681	682.021	10896	1221900	0.891726	0.055816
Mali	641	641.961	10256	1240190	0.826970	0.051763
Saudi Arabia	875	876.312	14000	2149690	0.651256	0.040764
Venezuela	362.6	363.143	5801.6	912050	0.636105	0.039816
Lesotho	1.6	1.6024	25.6	30350	0.084349	0.005279

Track density

To look at the relationship between area and rail network coverage, re-order Table 7.1 and arrange it by track density (Table 7.2). The density index shows that mid-sized industrial countries have extensive rail development. When the continental measures are mixed with the country measures, they offer additional insight that characteristically comes from shifting scale: the density of the railway network in the United States is about half that as for the entire European continent.

The order based on density fits with general knowledge of the world. The anomaly in Cuba comes from the presence of narrow and multiple gauge track induced by the sugar cane industry for hauling the crop out from the fields. Others, such as in the banana republics, arise once again from additional narrow gauge track needed to speed the fresh-cut bananas from the plantations into the 55 degree (F) holds of ships. From a more general policy viewpoint, data such as that in Table 7.2 might be used as support for legislation promoting the extension of rail networks in larger industrialized nations (such as the United States of America). Indeed, the lessons from the past show (even early in this century) that the Western European arrangement of rail development, in mid-sized political units, couples nations one after another into a more global rail network serving an entire subcontinent. The data of Table 7.2 suggests that this lesson from the past is one that has endured: in the United States (among others), the national rail network boasts nowhere near the same degree of density of land coverage as do those in Western Europe.

These are not new observations; Ginsburg and Berry presented extensive work-ups on railway densities for many of the world's nations through the mid-1950s and they offered the reader a carefully thought out context in which to place their observations--many of which remain true today, over 30 years later (quotation below if from page 60).

> Accessibility is a major factor in the developmental equation. Elements in the resource endowment need to be accessible in order to be used productively. Interaction among people, and therefore the dissemination of ideas and the sharing of knowledge and skills, requires a high physical mobility of people. More important, perhaps, the size of market in a given country and the degree to which its resources, both natural and otherwise, can be mobilized, is reflected in the size and quality of its transportation system. ... There are at least three ways to show the density of a railway network. The first relates railway kilometrage to area; the second relates railways to numbers of people; and the third attempts to combine the qualities of the first two.

The maps of Railway Density in Ginsburg and Berry are choropleth maps displaying different classes of density percentages. This strategy presents more quantitative information than does that of Jefferson, but the maps of Jefferson create a simpler, and easier to read, visual display. Current technology, involving computerized databases, spreadsheets (for numerical analysis), and GISs (for spatial analysis and display) can all be used on personal computing equipment. We illustrate how a map of railway density, similar to one in Ginsburg and Berry, might easily be updated using current data (as a simple map). From Jefferson to Ginsburg and Berry to current data sets, the interested reader might find it a useful challenge to update the Ginsburg and Berry Atlas, or other classical materials, using current technology.

Graphs of railway density
Assign numbers, beginning with the number 1--in an increasing fashion, to the lines in the data set of Table 7.2. Thus, partitioned Germany are numbers 1 and 2, Cuba is 3, Switzerland is 4, Belgium is 5, and so forth. These numbers are values that show rank within the set

of density values calculated using Jefferson buffers. When the column of ranking numbers is graphed on the x-axis (Figure 7.7) and the railway density values are graphed on the y-axis, the characteristic sharply decreasing curve often associated with rank-size phenomena emerges; Ginsburg and Berry note an almost identical curve using mid-1950s data. One might fit an exponential curve to it, but really there is no reason to do so. The pattern is displayed quite clearly by the actual curve and it would be meaningless to use this curve to interpolate or extrapolate the ranked countries. Curve fitting is a useful tool; it is not a universal tool.

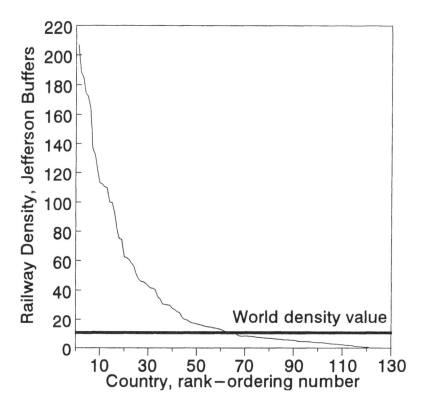

Figure 7.7. Railway density, in the manner of Ginsburg and Berry. Jefferson buffers based on 1989 data. World Jefferson density value=14.82771.

Figure 7.7 also has the World value for average railway density marked on it; in following the theme of comparing the U.S. railway density to that of other nations, it might be useful to insert the same line to the U.S.A. on another copy of this graph (Figure 7.8). A different visual display shows clearly, once again, that the density of the U.S. railnet, while greater than the world average is still far less than many other countries; this was the case in the 1920s, in the 1950s, and it remains the case today.

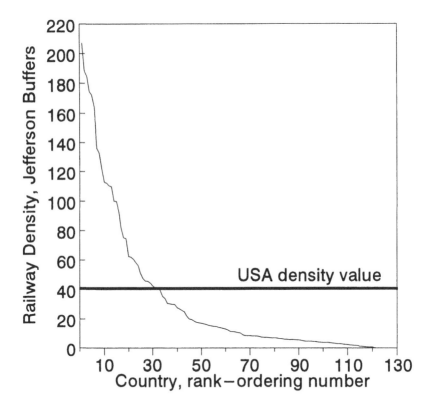

Figure 7.8. Railway density, in the manner of Ginsburg and Berry. Jefferson buffers based on 1989 data. U.S.A. Jefferson density value=41.27398.

Fitting maps to curves

When Table 7.2 is saved in a worksheet by itself in Lotus 1-2-3, it is a straightforward task to then import it into MapInfo and append it to the underlying database in that GIS. Once this is done, the data can be mapped and shaded in classes chosen by the user. Figure 7.9 shows a mapped version of Figure 7.7--railway densities that are above the World average are colored black and those below it are gray. The advantage the map has over the graph is that one can see which countries fall above and below the average. The advantage the graph has over the map is that numerical position of each country is shown.

Fortunately, the numerical and spatial software have a clear interface, so that users can generate both pictures. The only difficulty comes insofar as some data may be lost in moving a database file from the spreadsheet to the GIS; this happens, for example, when country names are not the same in the two data bases; for example, the MapInfo data base refers to a country in western Africa as "Upper Volta" whereas the World Resources Institute data base calls it "Burkina Faso." Thus, entries that have name discrepancies, as well as those for which there is partitioned data in one spreadsheet and non-partitioned data in the other (such as for Germany and various parts of the former Soviet Union), are lost in the transfer. Some of this difficulty can of course be corrected by entering new names for old by hand. We have left the map in Figure 7.9 exactly as it was produced, with no cleaning of the databases to maximize the mesh. The reason for doing so, is that the data loss is indicative of where there might be spots of boundary turmoil in the world--the electronic interface is itself a surrogate for political unrest.

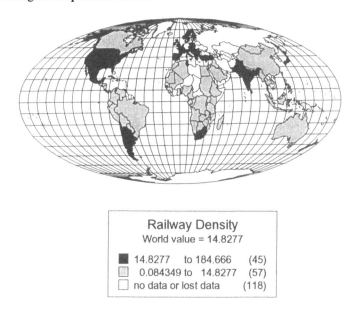

Figure 7.9. Railway density map partitioned by Jefferson world density value.

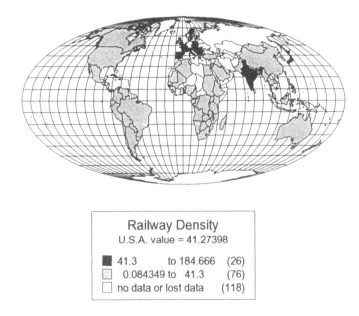

Railway Density
U.S.A. value = 41.27398

■ 41.3	to 184.666	(26)
☐ 0.084349 to	41.3	(76)
☐ no data or lost data		(118)

Figure 7.10. Railway density map partitioned by Jefferson world density value.

In a similar manner, the graph in Figure 7.8 can be realized as a map (Figure 7.10). On it one can see quite clearly which particular countries outrank the U.S.A. in railway density. Once the data set has been imported into the GIS, it becomes a matter of mere seconds or minutes to create any of a variety of maps. The time spent learning to use the software is well worth it.

References

1. Ginsburg, N. S. with an Appendix by Berry, B. J. L.. *Atlas of Economic Development, Chicago*, University of Chicago Press, 1961.
2. Jefferson, M. The civilizing rails. *Economic Geography*, 1928, 4, 217-231.
3. Kansky, K. J. *Structure of Transportation Networks: Relationships between Network Geometry and Regional Characteristics*, Research Paper no. 84, Chicago, University of Chicago, Department of Geography, 1963.

CHAPTER 8

ENVIRONMENTAL TOXICITY DATA ANALYSIS

ANALYTICAL TECHNIQUES/TOOLS USED:

Straight line curve-fitting—least squares
Fitting curves to maps

DATA TYPE: ABUNDANT DATA OF VARIABLE QUALITY

cleaning, analysis, and graphing of data

Overview of Data

Environmental toxicity is a recent term that refers to the complex interactions of global atmospheric conditions, local air pollution, surface water, ground water, and solid waste, among others. Typically, there are initially low levels of industrial or agricultural production coupled with low levels of environmental toxicity. When production and population increase, associated toxic products also increase. Eventually the increase reaches a level that is not acceptable to the general public, and the consequent outcry creates a demand to control pollution (Ness, Drake, and Brechin 1993).

Data sets concerning components of environmental toxicity are available in a number of data bases; however, their quality and extent of coverage is highly variable. Because toxicity is simultaneously a global and a local phenomenon, it is often unclear what information can be derived by comparing, interpolating, or extrapolating from data sets. Thus, it is important to limit what one looks at and to consider it from a variety of perspectives.

With any data set (presented in electronic or paper format), it is important first to examine the set for interesting or unusual patterns in the display. These patterns often influence decisions in choosing subsets of data and tools to analyze subsets.

PATTERNS IN DATA--WHAT TO LOOK FOR

1. What is the general organizational scheme of the entire set? Is it arranged alphabetically, numerically, or in some other fashion?

2. Are the real-world entries in the Table (nations, states, counties) expressed as comparable units? For example, county data and national data are generally not comparable.

3. Are the numerical entries in the Table expressed in comparable units? For example, data in one column might measure percentages while data in another column might measure thousands of dollars--these columns would not be comparable.

4. Are there gaps in the data? If so, what is their significance to the questions you wish to have the data answer?

The World Resources Institute data base has over 500 variables; probably more than one-third of them might be easily tied to the notion of environmental toxicity--from the variables having to do with waste products, to those having to do with soil degradation, to those involving toxic substances, to a host of others. One of the difficulties with this topic is knowing how to begin. A starting point is to look at a number of the variables with the framework in mind that is suggested in points 1 to 4 above.

1. What is the general organizational scheme of the entire set? Is it arranged alphabetically, numerically, or in some other fashion?

The data sets can all be sorted either alphabetically or numerically.

2. Are the real-world entries in the Table (nations, states, counties) expressed as comparable units? For example, county data and national data are generally not comparable.

In most data bases there are entries for nations and for continents; while these units are not directly comparable, it may be of particular importance in considering environmental toxicity to have both global and local scales of data present within a single spreadsheet. Continental statistics can be used as standards against which to measure national data.

3. Are the numerical entries in the Table expressed in comparable units? For example, data in one column might measure percentages while data in another column might measure thousands of dollars--these columns would not be comparable.

There are many different units in the many different tables; they are all units within the metric system.

4. Are there gaps in the data? If so, what is their significance to the questions you wish to have the data answer?

There are many gaps in the data in the tables related to environmental toxicity. Evidently there is considerable unevenness in the collection of data; there are many variables reporting only one year, or a small number of years, of statistics.

Limiting data choices

We have chosen to focus on the problem of limiting data selection by choosing comparable regions from different parts of the world: what is true for both regions on one variable may not be true for both on another variable. The variation may reflect policy differences. To begin, we considered the variable involving the production of sulfur dioxide in thousands of tons. Sulfur dioxide is a particularly toxic by-product that comes from burning low grade coal. Both the United States and countries in Central Europe are traditional users of coal; some are moving away from heavy use of coal, others are not.

Generally, it seems inappropriate to compare any of the countries of Central Europe, as a single political unit, to the United States. Environmental toxicity is too heavily tied to river systems, climatic regimes, and other broader geographic patterns to limit it to small political units. One natural unit that is good for making environmental comparisons involving drainage (directly or indirectly) is that of the river basin. Thus, we compare aggregate data for the countries that lie in the drainage basin of the Danube River to that of the United States. Any country that contains a tributary of the Danube River, visible on a map of the region on which 1 inch represents 64 miles, was included in the drainage basin. Albania, Italy, and Switzerland are at the tips of the drainage tree; Germany, Austria, Czechoslovakia, Hungary, Romania, and Bulgaria are all directly adjacent to the main stream of the river.

Data on environmental toxicity--comparisons of disparate regions

Table 8.1 shows data for the Danube drainage basin and the United States concerning sulfur dioxide (SO2) emissions. It does so for three years: 1980, 1985, and 1989. Scattered data for other years was available within the data base; one of the reasons for limiting the view is to get a solid base of data. Thus, only those years for which there was complete data were included in this table. An immediate result that follows from the table is that the total emissions for the Danube basin, and for the United States are quite similar when compared directly as totals. The sequence of totals is declining in both cases, as time goes forward, suggesting some improvement. One of the problems with data on this topic is, that because the problems associated with toxicity cut across geographic scales, one can often argue both sides of the same topic using the same table. Thus, when one focuses on the totals from the first three numerical columns, a nice aerial view of the Danube drainage basin, showing heavy industry, a not-so-beautiful Brown Danube, and dense smog, and then shows the similar totals, an argument can be made that the U.S. is in a pollution situation close to that of the Danube basin. On the other hand, when land is factored into the equation, the concentrations of SO2 per hectare are much higher in the Danube basin; they are 4 to 5 times as high as in the U.S. A reasonable question to pursue from here might be to consider whether total volume is more important that local concentration. Environmental toxicity data can be very difficult to interpret.

TABLE 8.1 (source: World Resources Institute)

Sulfur dioxide − − thousand tons
Danube River basin and U.S.A.

	1980	1985	1989	1989 Land Area Thous. Ha.	1989 SO2/area times 100
Albania	50	50	50	2875	1.739130
Austria	346	158	124	8385	1.478831
Bulgaria	1034	1140	1030	11091	9.286809
Czechoslovakia	3100	3150	2800	12787	21.89723
German Dem Re	5000	5000	5210	10833	48.09378
Germany, Fed R	3200	2400	1500	24858	6.034274
Hungary	1634	1420	1218	9303	13.09255
Italy	3800	2504	2410	30127	7.999468
Poland	4100	4300	3910	31268	12.50479
Romania	200	200	200	23750	0.842105
Switzerland	126	96	74	4129	1.792201
Yugoslavia	1176	1500	1650	25580	6.450351
Basin total	23766	21918	20176	194986	10.34740
United States	23400	21100	20700	937261	2.208563

Given the close fit on total SO2 emissions between the two geographic units, one might then be tempted to investigate other variables related to environmental toxicity for these two regions. Table 8.2 shows the results of using World Resources Institute data concerning fertilizer consumption for the two regions. Again, the U.S. and the Danube basin use similar total amounts; in this case, however, both are increasing sequences over time, and the time interval for which data was available is far more extensive than it was in the case of SO2. Because the U.S. has greater extent than the Danube basin, the figures per unit of land are of course much lower for the U.S. than they are for the Danube basin. Again, though, if only the totals are presented one might paint a far different picture than if both totals and concentrations per land units are presented simultaneously.

Another toxicity variable with an extensive time series in the data base involves industrial CO2 emissions. Table 8.3 shows data for total industrial CO2 emissions; the U.S. is about twice as high on the totals as the Danube basin (but of course lower per unit of land). Topographic and climatic concerns, such as positions of factories relative to mountains and the prevailing winds, can be very important in discussing the extent of difficulty that comes from emissions in local settings.

It seems worthwhile to look at a few other variables related to toxicity: methane emissions, CFC emissions (Table 8.4) and nitrogen oxide emissions (Table 8.5). Each of these shows either the situation that the US totals are similar to the Danube basin totals or that the US totals are much larger than the Danube totals. It is useful to look at a number of variables to note the broad patterns that emerge. There are a number of reasons that this pattern might occur, including the idea that the Danube basin countries might have policy controls on certain variables that are not in place in the US, or that demand for certain products (such as refrigeration) in the US is higher than it is in Danube countries.

Given the known high degree of pollution in many of the Danube countries, it might be useful to see if they have yet reached a stage where public outcry will force the hand of policy makers to act on behalf of improved public health. Thus, we next view the Danube region within the larger context of the entire set of countries for which there is data on S02 emissions.

TABLE 8.2 (source: World Resources Institute)

Fertilizer consumed — — thousand hectares
Danube River basin and U.S.A.

	1970	1971	1972	1973	1974	1975	1976	1977	1978	1979	1980	1981	1982
Albania	44	54	56	58	64	63	72	79	100	98	94	81	100
Austria	408	440	414	414	338	313	372	376	392	403	407	395	353
Bulgaria	639	636	646	633	567	679	658	742	738	820	830	1044	1038
Czechoslovakia	1282	1373	1417	1406	1549	1684	1675	1634	1749	1745	1730	1720	1742
German Dem Re	1535	1601	1618	1755	1840	1826	1804	1670	1670	1713	1637	1726	1408
Germany, Fed R	3228	3300	3239	3181	3248	3107	3406	3381	3439	3597	3532	3131	3246
Hungary	837	954	1017	1202	1336	1518	1388	1511	1539	1502	1399	1485	1528
Italy	1338	1450	1541	1412	1272	1490	1435	1795	2239	2355	2111	2057	2004
Poland	2572	2888	3047	3343	3460	3671	3586	3606	3567	3635	3499	3346	3182
Romania	594	633	639	854	921	1197	1144	1130	1480	1431	1223	1618	1409
Switzerland	148	151	138	159	145	143	152	156	176	184	181	169	170
Yugoslavia	631	669	716	706	674	720	738	802	855	870	824	1010	940
Basin total	13256	14149	14488	15123	15414	16411	16430	16882	17944	18353	17467	17782	17120
United States	15535	15580	16322	17516	15941	18914	20059	18676	20471	20941	21480	19439	16416

Fertilizer consumed -- thousand hectares
Danube River basin and U.S.A.

	1983	1984	1985	1986	1987	1988	1989	1989 Land Area Thous. Ha.	1989 Fert/area times 100
Albania	103	94	94	96	95	97	107	2875	3.721739130
Austria	382	390	388	316	334	322	308	8385	3.673225998
Bulgaria	1009	937	864	808	745	918	807	11091	7.276169867
Czechoslovakia	1776	1749	1734	1703	1556	1606	1641	12787	12.83334636
German Dem Re	1451	1566	1639	1642	1664	1805	1718	10833	15.85894950
Germany, Fed R	3136	3172	3185	3193	3146	3071	2873	24858	11.55764743
Hungary	1586	1524	1338	1383	1373	1418	1302	9303	13.99548532
Italy	2064	2091	2102	2061	2303	2092	1813	30127	6.017857735
Poland	3424	3280	3413	3583	3277	3625	3029	31268	9.687220161
Romania	1468	1506	1374	1386	1390	1426	1378	23750	5.802105263
Switzerland	177	180	180	173	177	178	176	4129	4.262533301
Yugoslavia	920	976	992	1023	1031	1015	897	25580	3.506645817
Basin total	17496	17465	17303	17367	17091	17573	16049	194986	8.230847342
United States	19768	19688	17831	17286	17792	17733	18709	937261	1.996135548

Industrial CO2 emissions – – thousand tons
Danube River basin and U.S.A.

TABLE 8.3 (source: World Resources Institute)

	1970	1971	1972	1973	1974	1975	1976	1977	1978	1979	1980	1981	1982
Albania	3414.84	3803.23	4422.44	3902.16	4118.33	6562.22	4590.99	5103.95	5232.19	8621.39	9669.29	10310.4	11252.1
Austria	50303.0	51746.6	55733.1	59646.2	56846.9	54146.5	58220.9	56007.9	57037.4	61056.8	52201.0	56055.5	53802.1
Bulgaria	79790.9	81941.6	84499.1	87419.3	89273.3	94238.0	92545.3	95011.1	100987.	106431.	111136.	110330.	115188.
Czechoslovakia	199105.	208723.	210203.	211083.	213607.	223712.	232407.	239552.	246748.	240607.	242168.	240145.	237965.
German Dem Re	270568.	275261.	272916.	276837.	277005.	278229.	285293.	292629.	295853.	302060.	306596.	307735.	308292.
Germany, Fed R	735943.	741916.	747888.	783520.	756883.	697790.	774925.	729637.	750515.	789936.	762188.	710577.	675982.
Hungary	72012.2	69872.4	70726.1	74445.1	75654.2	77079.5	81850.0	84389.2	87811.4	88390.3	85305.2	85224.6	86371.4
Italy	285627.	296527.	310985.	337000.	342737.	324740.	352374.	343122.	359057.	376552.	372130.	365494.	354708.
Poland	303316.	312117.	328598.	333944.	344218.	372833.	396327.	416153.	430765.	441522.	459597.	406136.	418326.
Romania	120355.	126712.	132852.	145570.	151356.	162978.	174629.	178271.	194067.	196214.	200772.	199325.	196844.
Switzerland	39497.9	41150.3	42198.2	45481.2	40758.3	38395.0	39963.2	40516.5	41641.3	39226.7	40923.2	38647.8	36500.7
Yugoslavia	68209.0	74624.6	71854.7	84671.3	83432.9	86583.9	89592.1	89339.3	94007.2	101998.	107007.	114503.	104376.
Basin total	2228144	2284397	2332879	2443521	2435893	2417291	2582720	2569735	2663724	2752620	2749696	2644488	2599611
United States	4270307	4298758	4496995	4672937	4510743	4319954	4624026	4652312	4739116	4766556	4614005	4435660	4202256

Industrial CO2 emissions -- thousand tons
Danube River basin and U.S.A.

	1983	1984	1985	1986	1987	1988	1989	1989 Land Area Thous. Ha.	1989 CO2/area times 100
Albania	11332.7	9284.57	9801.2	9724.25	9727.92	9911.12	9731.58	2875	338.4898782
Austria	51581.7	537728	53915.7	53215.9	53948.7	51706.3	51699.0	8385	616.5657722
Bulgaria	115910.	106266.	107406.	112477.	112917.	105471.	106988.	11091	964.6452078
Czechoslovakia	238581.	243930.	241813.	241135.	239420.	233404.	226347.	12787	1770.135794
German Dem Re	306020.	315155.	330551.	333874.	333570.	326946.	322629.	10833	2978.213385
Germany, Fed R.	672765.	685135.	678668.	677752.	663477.	669464.	641397.	24858	2580.247228
Hungary	89280.6	83007.9	83238.7	83685.7	81447.0	73866.2	64076.0	9303	688.7674083
Italy	342770.	350769.	357789.	348065.	363384.	371386.	389747.	30127	1293.680114
Poland	419289.	437056.	448532.	455651.	468207.	458970.	440929.	31268	1410.161903
Romania	202076.	221122.	189502.	205238.	212402.	211574.	212193.	23750	893.4451873
Switzerland	39941.2	39087.5	39736.0	42051.7	40153.7	40483.5	39325.7	4129	952.4270283
Yugoslavia	112220.	119527.	121003.	126331.	124788.	132750.	132900.	25580	519.5488975
Basin total	2601773	2664116	2661958	2689203	2703445	2685935	2637966	194986	1352.900421
United States	4211320	4351051	4401522	4413441	4594022	4813906	4869005	937261	519.4930046

TABLE 8.4 (source: World Resources Institute)

1989 Methane emissions – – thousand tons Danube River basin and U.		Land area Thous. Ha.	Meth/area times 100	1989 CFC emissions – – thousand tons	Land area Thous. Ha.	CFC/area times 100
Albania	94.7923	2875	3.297125		2875	0
Austria	310	8385	3.697078	3.348204	8385	0.039930
Bulgaria	330	11091	2.975385	1.059646	11091	0.009554
Czechoslovakia	820	12787	6.412762	3.759604	12787	0.029401
German Dem Re	680	10833	6.277116	7.196485	10833	0.066431
Germany, Fed R	3000	24858	12.06854	26.90846	24858	0.108248
Hungary	410	9303	4.407180	3.051594	9303	0.032802
Italy	2000	30127	6.638563	24.96683	30127	0.082871
Poland	2500	31268	7.995394	5.475274	31268	0.017510
Romania	1500	23750	6.315789	2.108471	23750	0.008877
Switzerland	240	4129	5.812545	2.457534	4129	0.059518
Yugoslavia	750	25580	2.931896	3.947896	25580	0.015433
Basin total	12634.7	194986	6.479845	84.28000	194986	0.043223
United States	37000	937261	3.947673	130	937261	0.013870

TABLE 8.5 (source: World Resources Institute)

Nitrogen oxide emissions – – thousand tons Danube River basin and U.S.A.	1985	1989	1989 Land area Thous. Ha.	NOx/area times 100
Albania	9	9	2875	0.313043
Austria	219	211	8385	2.516398
Bulgaria	150	150	11091	1.352447
Czechoslovakia	1127	950	12787	7.429420
German Dem Re	955	708	10833	6.535585
Germany, Fed R	2950	3000	24858	12.06854
Hungary	300	259	9303	2.784048
Italy	1595	1700	30127	5.642778
Poland	1500	1480	31268	4.733273
Romania	390	390	23750	1.642105
Switzerland	214	194	4129	4.698474
Yugoslavia	190	190	25580	0.742767
Basin total	9599	9241	194986	4.739314
United States	19800	19800	937261	2.112538

Straight line curve fitting--least squares analysis--Danube basin

Table 8.6 shows the data for SO2 for the Danube drainage basin, the U.S., and all the other entries that were available within the World Resources Institute data base. The entries in Table 8.6 are sorted according to 1989 emissions levels in decreasing order (Column I). In Column C, each country is assigned a rank based on the 1989 ordering. In Column B, each country is assigned a rank based on ordering the emission data in descending order using the 1985 data in column H. One facet of toxicity data that can be measured is whether or not there is a reported reduction in the emissions level over time. Rank-ordering makes this sort of assessment an easy matter. In the five year time span, Austria and Sweden have both dropped four positions on the list-- their performance in reducing SO2 emissions has improved, at the very least, relative to this set of countries. Indeed, both also improved absolutely in reducing their values. A least squares line (Table 8.6) fit to the scatter of dots (Figure 8.1) will measure the degree of association between the two rank orderings.

STRAIGHT LINE FIT TO THE DATA OF TABLE 8.6
(Refer to Table 8.6 and Figure 8.1)

1. Enter the values of the independent variable, 1985 data-- Column C in Table 8.6.

2. Enter the values of the dependent variable--Column B in Table 8.6.

3. Choose the regression feature from the software, with the x values as in step 1 and the y values as in step 2.

4. Choose the output range as a blank area in the spreadsheet. Then proceed with the calculation as directed by the software; the output from the regression will appear in a form similar to the one in Table 8.6, bottom half (produced in Lotus 1-2-3, release. 2.3).

5. The equation below the output range must generally be derived by the user from the regression output. The slope-intercept form for the equation of a straight line (y=mx+b) is used. The "X Coefficient" from the regression output is used as "m". The "Constant" from the regression output is used as b. Thus, the least squares equation that fits the data can be read directly from the regression output.

TABLE 8.6 (source: World Resources Institute)
Sulfur dioxide – – thousand tons

Diff.	89	85		1970	1975	1980	1985	1989	Lin. est.	Res.
0	1	1	United States	28400	25900	23400	21100	20700	1.433	-0.43
0	2	2	U.S.S.R.			12800	11100	9318	2.401	-0.40
0	3	3	German Dem Rep			5000	5000	5210	3.369	-0.36
1	5	4	United Kingdom	6424	5370	4848	3676	3552	4.337	0.663
-1	4	5	Poland			4100	4300	3910	5.305	-1.30
2	8	6	Italy	2830	3331	3800	2504	2410	6.273	1.727
3	10	7	France	2966	3328	3510	1846	1520	7.241	2.759
-2	6	8	Spain		3003	3250	3250	3250	8.209	-2.20
2	11	9	Germany, Fed R	3743	3334	3200	2400	1500	9.177	1.823
-3	7	10	Czechoslovakia			3100	3150	2800	10.14	-3.14
1	12	11	Hungary			1634	1420	1218	11.11	0.887
-3	9	12	Yugoslavia			1176	1500	1650	12.08	-3.08
0	13	13	Bulgaria			1034	1140	1030	13.04	-0.04
0	14	14	Belgium			828	452	414	14.01	-0.01
2	17	15	Finland	515	535	584	372	318	14.98	2.015
4	20	16	Sweden	930	690	502	270	220	15.95	4.047
1	18	17	Netherlands	807	429	464	276	290	16.92	1.079
1	19	18	Denmark	574	418	450	340	242	17.88	1.111
-4	15	19	Greece			400	360	360	18.85	-3.85
4	24	20	Austria			346	158	124	19.82	4.175
-5	16	21	Turkey			276	322	354	20.79	-4.79
-1	21	22	Portugal	116	178	266	204	204	21.76	-0.76
0	23	23	Ireland			220	138	148	22.72	0.271
-2	22	24	Romania			200	200	200	23.69	-1.69
0	25	25	Norway	171	137	142	98	74	24.66	0.335
0	26	26	Switzerland	125	109	126	96	74	25.63	0.367
0	27	27	Albania			50	50	50	26.60	0.399
0	28	28	Luxembourg			24	16	12	27.56	0.431
0	29	29	Iceland			6	6	6	28.53	0.463
*		5	Canada	6677	5319	4643	3704			
*		13	Japan	4973	2586	1263				

Regression Output:

Constant	0.465
Std Err of Y Est	2.143
R Squared	0.938
No. of Observations	29
Degrees of Freedom	27

X Coefficient(s)	0.968
Std Err of Coef.	0.047

$y = 0.968 * x + 0.465$

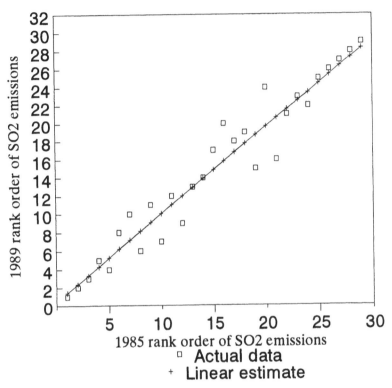

Figure 8.1. 1985 sulfur dioxide emission; Danube basin.

6. In a separate column of the spreadsheet, Column J in Table 8.6, enter the equation derived from the regression: in cell J5, enter the formula 0.968*C5+0.465. The value 1.433 should appear at the top of Column J. Then, copy the cell content from J5 to the cells below it; this should produce the numerical range shown in Column J of Table 8.6

7. Graph the results; select an XY-graph. Put the entries for the 1985 rank-ordering in the X-range, and for the 1989 rank-ordering in the A range. Enter the entries of Column J of Table 8.6 in the B range of the graph. The result should look like Figure 8.1.

BLACK BOX SUMMARY
see Introduction for theoretical explanation

LEAST SQUARES REGRESSION LINE

y=mx+b

where
m is the slope of the line, or the "x-coefficient"
b is the y-intercept, or the "constant."

The linear fit is quite good; the R-squared value is 0.93 and the standard errors of both the constant and the coefficient of x are relatively low, although the error in the Y-estimate (error for the constant) is about half of the maximum deviation from the line. The error terms suggest that the slope of the line fits the scatter quite well, but that some error in up-down placement might have occurred. Because the residual plot (Figure 8.2) shows no particular non-linear pattern, it suggests that a straight line was a reasonable choice to fit to this scatter. There are a number of questions that one might now consider, as a result of having fit a line to the scatter.

Is there any element of causation involved; the line measures association, at the very least, but it need not measure causation. In this case the independent variable is the 1985 rank ordering of SO2 emissions--does a particular rank value tend to cause the same, or nearly the same, rank value in 1989? It appears that indeed this is the case; when policy maintains the status quo, there is no reason for ranking to change. When policy is strengthened, there is deviation-- those points in the scatter above the line have improved their emissions policy (Austria, Sweden, France, Finland, and others); those lower than the line have dropped in their control of SO2 emissions. Some of the countries below the line are former USSR countries; perhaps the opportunity afforded by new-found freedom has outstripped environmental controls.

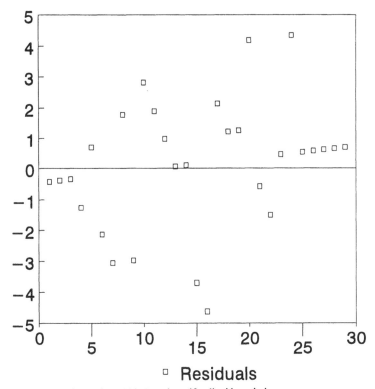

Figure 8.2. Residual plot, 1985. Danube sulfur dioxide emissions.

Spatial cut-offs suggest improvements

In total volume of emissions, the U.S. is similar to the entire Danube region; in that regard, it captures the value for a single political unit that might then be assigned to each individual political unit--to do so, however, would be to commit the ecological fallacy--that of assigning the characteristics of the whole to each of its parts. What can be done however, is to use the U.S. value, as one at a different scale (in terms of area), to partition the countries of the Danube region into those above and below the U.S. emissions total for 1989 (Figure 8.3).

When this change in scale is performed, Austria, in the heart of the Danube basin, stands out. Indeed, the evidence of maps supports the evidence of the tables and suggests that one piece of reliable information is that Austria has made significant strides in reducing SO2 emissions; perhaps the U.S. can look to that part of the Danube basin, rather than to the grim slide show depicted earlier in this chapter, to study policies in the management of SO2 toxic emissions.

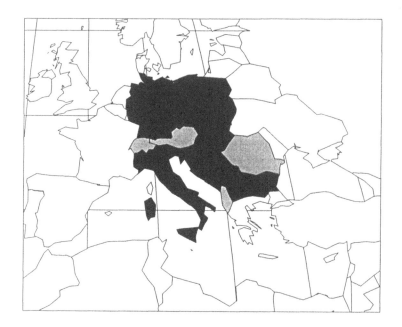

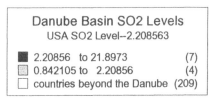

Figure 8.3. Danube basin partitioned by USA sulfur dioxide level.

References

1. Land, T. The Danube: dams over troubled waters. *Nature*. Jan. 23, 1992.
2. Meadows, D. H., Meadows, D. L., Randers, J. *Beyond the Limits*, Chelsea Green Publishing Company, Post Mills, Vermont, 1992.
3. Ness, G., Drake, W., and Brechin, S. Population-Environment Dynamics: Ideas and Observations, The University of Michigan Press, Ann Arbor, Michigan, 1993.

CHAPTER 9

URBANIZATION DATA ANALYSIS

ANALYTICAL TECHNIQUES/TOOLS USED:

Straight line curve-fitting--least squares
Partitioning of data: scale transformation.
Residual plots
Examination and removal of outliers

DATA TYPE: ABUNDANT AND BASELINE DATA

cleaning, analysis, and graphing of data

Overview of Data

Occasionally one notes the presence of a new variable in a data base that seems particularly striking--why is it there; to what might it be related? Such was the case with the variable "Botanical Gardens" within the World Resources Institute data base. One might imagine botanical gardens as a surrogate measure for extent of urbanization. Cities draw life from resources (energy) and typically ideas are created through the interaction made possible by the concentration of resources. Thus, one expects greater concentrations of numbers of patents, libraries, museums, airports, zoological parks, and botanical gardens in very large cities with an active urban metabolism.

The data set concerning botanical gardens has entries only for the year 1991; it is not typical of the traditional data sets involving urbanization, such as per capita income data, energy use data, telephone data or municipal waste generation data, many of which have time series covering 20 years or more. We investigate the possibility of using this intriguing new data set in conjunction with a more traditional one, of number of people in a country living in cities of greater than one million population, to see the extent to which the number of botanical gardens is related to extent of urbanization.

TABLE 9.1 (source: World Resources Institute)

	Bot. Gardens Number		People in cities Over 1 million thousands
	1991		1990
China	66	144.517	103994
United States	247	126.091	90345
India	68	102.685	73007
Brazil	11	75.6567	52986
U.S.S.R.	160	63.9158	44289
Japan	59	49.8839	33895
Mexico	30	42.9813	28782
Korea, Rep	5	32.9387	21343
Indonesia	5	28.9872	18416
Pakistan	5	26.3615	16471
Italy	48	23.6534	14465
Argentina	9	22.6598	13729
United Kingdom	60	22.2251	13407
Egypt	5	21.3030	12724
Iran, Islamic Rep	3	21.3003	12722
Turkey	6	20.4269	12075
France	66	18.6503	10759
Nigeria	5	16.3188	9032
Colombia	13	16.2891	9010
Bangladesh	2	16.1879	8935
Spain	8	15.8274	8668
Australia	60	15.6465	8534
Philippines	9	15.5669	8475
Germany, Fed Rep	59	14.8473	7942
Canada	18	14.7798	7892
Thailand	5	13.7862	7156
Poland	25	13.2057	6726
Peru	5	12.5591	6247
Venezuela	7	11.2023	5242
Chile	9	10.5165	4734
Viet Nam	2	10.0035	4354
Morocco	2	9.90502	4281
Iraq	1	9.58507	4044
Syrian Arab Rep	0	9.22327	3776
Zaire	2	8.85742	3505
Greece	4	8.77102	3441
Myanmar	2	8.57392	3295
Saudi Arabia	2	8.43352	3191
Algeria	3	8.22022	3033
Ecuador	2	8.06092	2915
Singapore	1	7.80172	2723
Romania	10	7.08757	2194
Cote d'Ivoire	1	7.05247	2168

Hungary	17	6.98092	2115
Cuba	8	6.95932	2099
Netherlands	39	6.95932	2099
Austria	11	6.95527	2096
Libya	1	6.90937	2062
Sudan	1	6.75412	1947
Ethiopia	1	6.67852	1891
Israel	7	6.66772	1883
Angola	1	6.44362	1717
Malaysia	9	6.43552	1711
Sweden	9	6.36937	1662
Tanzania	2	6.36262	1657
Tunisia	1	6.33427	1636
Portugal	6	6.28972	1603
Mozambique	2	6.26947	1588
Yugoslavia	32	6.25192	1575
Afghanistan	0	6.23842	1565
Kenya	5	6.15472	1503
Senegal	3	6.13987	1492
Denmark	7	5.99542	1385
Guinea	0	5.87527	1296
Czechoslovakia	34	5.87257	1294
German Dem Rep	14	5.84152	1271
Bolivia	3	5.79157	1234
Uruguay	1	5.74162	1197
Bulgaria	9	5.73217	1190
Ghana	3	5.61202	1101
Kuwait	0	5.58232	1079
Jordan	0	5.50942	1025
Costa Rica	2	5.49727	1016
Nicaragua	1	5.49187	1012
Finland	8	5.48782	1009

Regression Output:

Constant	4.12567
Std Err of Y Est	26.4540
R Squared	0.48899
No. of Observations	75
Degrees of Freedom	73

X Coefficient(s)	0.00135
Std Err of Coef.	0.00016

Bot Gar=f(pop 1 mil)
y=0.00135*x+4.12567

With any data set (presented in electronic or paper format), it is important first to examine the set for interesting or unusual patterns in the display. These patterns often influence decisions in choosing subsets of data and tools to analyze subsets.

PATTERNS IN DATA--WHAT TO LOOK FOR

1. What is the general organizational scheme of the entire set? Is it arranged alphabetically, numerically, or in some other fashion?

2. Are the real-world entries in the Table (nations, states, counties) expressed as comparable units? For example, county data and national data are generally not comparable.

3. Are the numerical entries in the Table expressed in comparable units? For example, data in one column might measure percentages while data in another column might measure thousands of dollars--these columns would not be comparable.

4. Are there gaps in the data? If so, what is their significance to the questions you wish to have the data answer?

These four points were addressed in the creation of Table 9.1.

1. What is the general organizational scheme of the entire set? Is it arranged alphabetically, numerically, or in some other fashion?

The organizational scheme of the data on botanical gardens fits with that of the other tables in the data base, making it compatible for use with them; all can be sorted by variable and then by country or vice-versa.

2. Are the real-world entries in the Table (nations, states, counties) expressed as comparable units? For example, county data and national data are generally not comparable.

The botanical gardens data had entries at both the country and the continent level as did the variable on number of people living in cities of more than a million. The two sets were compatible in this regard.

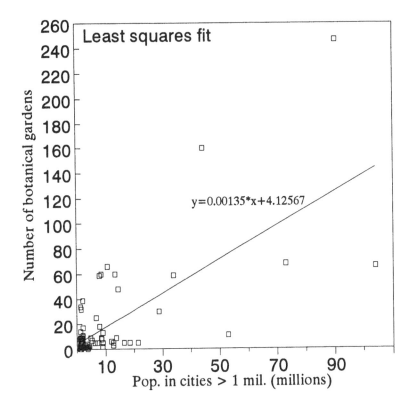

Figure 9.1. Global least squares line--botanical gardens and urban data.

3.   Are the numerical entries in the Table expressed in comparable units? For example, data in one column might measure percentages while data in another column might measure thousands of dollars--these columns would not be comparable.

The units for botanical gardens are number of gardens per country or continent; the units for people in cities of over 1 million is given in thousands of people. Thus, the values along the x-axis in Figure 9.1 are in 90 (thousand) units of the units in Table 9.1--a thousand thousand is a million, so the units along the x-axis are millions. For example, in the U.S.A., over 90 million people live in cities of over 1 million.

4. Are there gaps in the data? If so, what is their significance to the questions you wish to have the data answer?

There are many gaps; the data set concerning botanical gardens contain far fewer entries than do many of the data sets concerning urban phenomena. The botanical gardens data set is for one year only, 1991; the most recent data in most of the other data sets is for 1989 or 1990. Thus, we compare on the basis of "most recent" rather than on the basis of matching exact years.

Straight line curve fitting--least squares analysis--gardens and cities

One way to test the idea of association between population concentration and number of botanical gardens is to fit a line of least squares, using population as the independent variable and botanical gardens as the dependent variable. Logically, it would seem that a certain threshold population needs to be present in order to justify the expense of constructing and maintaining a botanical garden; it is not clear, though, that a great many more people beyond the threshold necessarily means that a great many more botanical gardens will be built.

In Table 9.1, Column D displays the values for the number of people in cities over 1 million (in thousands) by country and Column B displays the number of botanical gardens by country.

STRAIGHT LINE FIT TO THE DATA OF TABLE 9.1
(Refer to Table 9.1 and Figure 9.1)

1. Enter the values of the independent variable--Column D in Table 9.1.

2. Enter the values of the dependent variable--Column B in Table 9.1.

3. Choose the regression feature from the software, with the x values as in step 1 and the y values as in step 2.

4. Choose the output range as a blank area in the spreadsheet. Then proceed with the calculation as directed by the software; the output from the regression will appear in a form similar to the one in Table 9.1, bottom half (produced in Lotus 1-2-3, release. 2.3).

5. The equation below the output range must generally be derived by the user from the regression output. The slope-intercept form for the equation of a straight line (y=mx+b) is used. The "X Coefficient" from the regression output is used as "m". The "Constant" from the regression output is used as b. Thus, the least squares equation that fits the data can be read directly from the regression output.

6. In a separate column of the spreadsheet, Column C in Table 9.1, enter the equation derived from the regression: in cell C6, enter the formula 0.00135*B6+4.12567. The value 144.517

should appear at the top of Column C. Then, copy the cell content from C6 to the cells below it; this should produce the numerical range shown in Column C of Table 9.1

7. Graph the results; select an XY-graph. Put the entries for the people in cities over 1 million in the X-range, and the number of botanical gardens values in the A range. Enter the entries of Column C of Table 9.1 in the B range of the graph. The result should look like Figure 9.1.

The R-squared value for this fit is not particularly good; it is 0.48. Of course, one might well not expect an association of 0.99--there are other factors associated with numbers of botanical gardens. It would not be prudent, therefore, to assert that there is a causal relationship between total numbers of people living in big cities and total numbers of botanical gardens. There appears to be some association, but there are evidently other variables that are associated with numbers of botanical gardens, too. The standard error of the x-estimate is low; that of the y-estimate is high. Thus, the slope of the line that is fit to the scatter of points in Figure 9.1 is a tight fit; however, there is much flexibility in the y-intercept. One would expect that in general, given a population value, that the corresponding number of gardens derived from using this least squares fit, would lie within 26 of the number forecast by the curve. This is certainly not a satisfactory situation; not that many countries even have 26 botanical gardens!

BLACK BOX SUMMARY
see Introduction for theoretical explanation

LEAST SQUARES REGRESSION LINE

$y=mx+b$

where
m is the slope of the line, or the "x-coefficient"
b is the y-intercept, or the "constant."

Partitioning of the data: least squares fits by continent

One might therefore discard the idea of trying to view number of botanical gardens as a function of number of people in cities of over 1 million. An alternate strategy is to partition the data in some way that seems reasonable from looking at its structure. In this case, note that many of the relatively small European countries have relatively large numbers of botanical gardens; perhaps it would make sense to partition the data by continent and then calculate least squares (or other) fits for each. Table 9.2 shows the results of executing a least squares fit for each of the continents. The linear equations used for representing numbers of botanical gardens as a function of the number of people in cities of over 1 million are shown in Table 9.3.

The R-squared values for Africa (.39) and for South America (.29) are substantially worse than was that for the entire distribution (.48). The R-squared values for Europe (.50) and Asia (.47) are about the same as for the entire distribution (.48). The R-squared value for North America (.95) was almost double that of the entire distribution (.48). One might consider that the association under consideration is one that shows the bias of observing the situation from North America. There are also changes in the y-intercept error: That for Africa (1.2) shows great improvement over that for the entire distribution (26), as does that for South America (3.7). The y-intercept errors for North and Central America (21), for Europe (15), and for Asia (27) are still so large as to make any sort of evaluation of a relationship between the variables highly tenuous. The graphs of these associations (Figures 9.2 to 9.6) display where there is good fit within the continental distributions, and where there is not.

The R-squared for South America is quite low; but, its y-intercept error is quite small. It may be worthwhile to see if the R-squared value can be improved; one way to do this is to remove any obvious outliers that seem to interfere with fitting the line to a pattern in the scatter diagram. A glance at Figure 9.3 therefore suggests removing the outlier with roughly 11 million living in big cities--Brazil. When this is done, and the regression recalculated, the result is shown in the last part of Table 9.3, and it is graphed in Figure 9.7. The R-squared value was indeed substantially improved from 0.29 to 0.55, as was the y-intercept error (from 3.75 to 2.97). Removing judiciously selected outliers can improve curves that are fit to a distribution. The effect of doing so is particularly noticeable when least squares are employed because the outliers contribute not just extra distance to the value, but the SQUARE of that distance. The contribution of outliers is too heavy in least squares fits, so removal of them will produce substantial improvement in fit.

TABLE 9.2 (source: World Resources Institute)

	Bot. Gardens Number		People in cities Over 1 million thousands	Residuals
	1991		1990	
WORLD	1553		782434	
AFRICA	82		58833	
1 Egypt	5	5.20036	12724	−0.2003
2 Nigeria	5	4.05584	9032	0.94416
3 Morocco	2	2.58303	4281	−0.5830
4 Zaire	2	2.34247	3505	−0.3424
5 Algeria	3	2.19615	3033	0.80385
6 Cote d'Ivoire	1	1.928	2168	−0.928
7 Libya	1	1.89514	2062	−0.8951
8 Sudan	1	1.85949	1947	−0.8594
9 Ethiopia	1	1.84213	1891	−0.8421
10 Angola	1	1.78819	1717	−0.7881
11 Tanzania	2	1.76959	1657	0.23041
12 Tunisia	1	1.76308	1636	−0.7630
13 Mozambique	2	1.7482	1588	0.2518
14 Kenya	5	1.72185	1503	3.27815
15 Senegal	3	1.71844	1492	1.28156
16 Guinea	0	1.65768	1296	−1.6576
17 Ghana	3	1.59723	1101	1.40277
SOUTH AMERICA	67		97294	
1 Brazil	11	12.0486	52986	−1.0486
2 Argentina	9	6.94525	13729	2.05475
3 Colombia	13	6.33178	9010	6.66822
4 Peru	5	5.97259	6247	−0.9725
5 Venezuela	7	5.84194	5242	1.15806
6 Chile	9	5.7759	4734	3.2241
7 Ecuador	2	5.53943	2915	−3.5394
8 Bolivia	3	5.3209	1234	−2.3209
9 Uruguay	1	5.31609	1197	−4.3160
NORTH & CEN AMER	355		135770	
1 United States	247	234.350	90345	12.6493
2 Mexico	30	69.3618	28782	−39.361
3 Canada	18	13.3766	7892	4.62334
4 Cuba	8	−2.1485	2099	10.1485
5 Costa Rica	2	−5.0510	1016	7.05102
6 Nicaragua	1	−5.0617	1012	6.06174
EUROPE	533		84901	
1 Italy	48	57.5890	14465	−9.5890
2 United Kingdom	60	54.0870	13407	5.91292
3 France	66	45.3222	10759	20.6778
4 Spain	8	38.4009	8668	−30.400
5 Germany, Fed Rep	59	35.9979	7942	23.0020
6 Poland	25	31.9729	6726	−6.9729

7 Greece	4	21.0996	3441	−17.099
8 Romania	10	16.9720	2194	−6.9720
9 Hungary	17	16.7105	2115	0.28944
10 Netherlands	39	16.6576	2099	22.3424
11 Austria	11	16.6476	2096	−5.6476
12 Sweden	9	15.2111	1662	−6.2111
13 Portugal	6	15.0158	1603	−9.0158
14 Yugoslavia	32	14.9231	1575	17.0768
15 Denmark	7	14.2942	1385	−7.2942
16 Czechoslovakia	34	13.9930	1294	20.0069
17 German Dem Rep	14	13.9169	1271	0.08308
18 Bulgaria	9	13.6488	1190	−4.6488
19 Finland	8	13.0497	1009	−5.0497
ASIA	271		352813	
1 China	66	105.068	103994	−39.068
2 India	68	74.0817	73007	−6.0817
3 U.S.S.R.	160	45.3637	44289	114.636
4 Japan	59	34.9697	33895	24.0302
5 Korea, Rep	5	22.4177	21343	−17.417
6 Indonesia	5	19.4907	18416	−14.490
7 Pakistan	5	17.5457	16471	−12.545
8 Iran, Islamic Rep	3	13.7967	12722	−10.796
9 Turkey	6	13.1497	12075	−7.1497
10 Bangladesh	2	10.0097	8935	−8.0097
11 Philippines	9	9.54971	8475	−0.5497
12 Thailand	5	8.23071	7156	−3.2307
13 Viet Nam	2	5.42871	4354	−3.4287
14 Iraq	1	5.11871	4044	−4.1187
15 Syrian Arab Rep	0	4.85071	3776	−4.8507
16 Myanmar	2	4.36971	3295	−2.3697
17 Saudi Arabia	2	4.26571	3191	−2.2657
18 Singapore	1	3.79771	2723	−2.7977
19 Israel	7	2.95771	1883	4.04229
20 Malaysia	9	2.78571	1711	6.21429
21 Afghanistan	0	2.63971	1565	−2.6397
22 Kuwait	0	2.15371	1079	−2.1537
23 Jordan	0	2.09971	1025	−2.0997
OCEANIA	85		8534	
1 Australia	60		8534	

TABLE 9.3

AFRICA

Regression Output:

Constant		1.25592
Std Err of Y Est		1.25318
R Squared		0.39688
No. of Observations		17
Degrees of Freedom		15
X Coefficient(s)	0.00031	
Std Err of Coef.	0.00010	

y=0.00031*x+1.25592

SOUTH AMERICA

Regression Output:

Constant		5.16048
Std Err of Y Est		3.75566
R Squared		0.29474
No. of Observations		9
Degrees of Freedom		7
X Coefficient(s)	0.00013	
Std Err of Coef.	0.00008	

y=0.00013*x+5.16048

NORTH AND CENTRAL AMERICA

Regression Output:

Constant		−7.7739
Std Err of Y Est		21.9058
R Squared		0.95889
No. of Observations		6
Degrees of Freedom		4
X Coefficient(s)	0.00268	
Std Err of Coef.	0.00027	

y=0.00268*x−7.7739

EUROPE

Regression Output:

Constant		9.70991
Std Err of Y Est		15.0306
R Squared		0.50345
No. of Observations		19
Degrees of Freedom		17
X Coefficient(s)	0.00331	
Std Err of Coef.	0.00079	

y=0.00331*x+9.70991

ASIA

	Regression Output:
Constant	1.07471
Std Err of Y Est	27.8911
R Squared	0.47088
No. of Observations	23
Degrees of Freedom	21
X Coefficient(s)	0.00100
Std Err of Coef.	0.00023

y=0.00100*x+1.07471

SOUTH AMERICA, LESS BRAZIL

	Regression Output:
Constant	2.10026
Std Err of Y Est	2.97378
R Squared	0.55364
No. of Observations	8
Degrees of Freedom	6
X Coefficient(s)	0.00072
Std Err of Coef.	0.00026

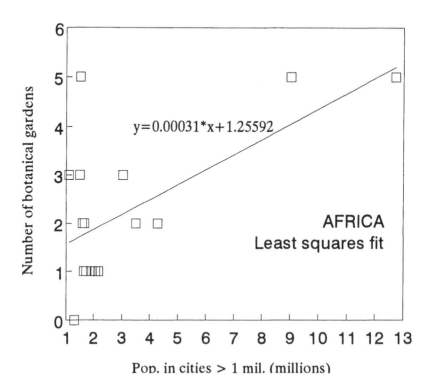

Figure 9.2. Africa, least squares fit. Botanical gardens and urban data.

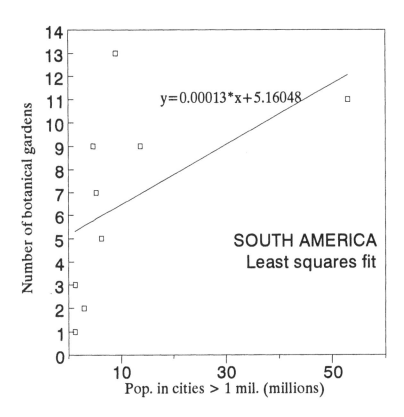

Figure 9.3. South America, least squares fit. Botanical gardens and urban data.

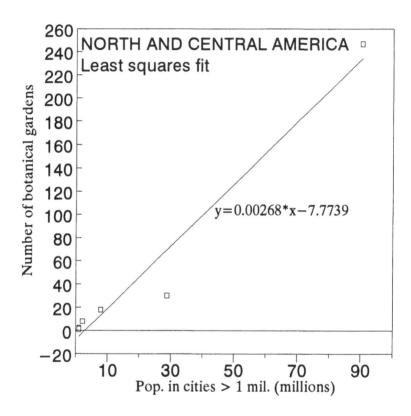

Figure 9.4. North and Central America, least squares fit. Botanical gardens and urban data.

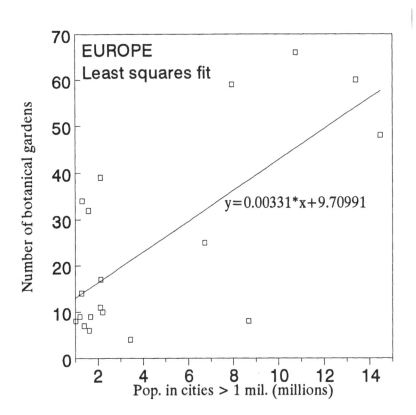

Figure 9.5. Europe, least squares fit. Botanical gardens and urban data.

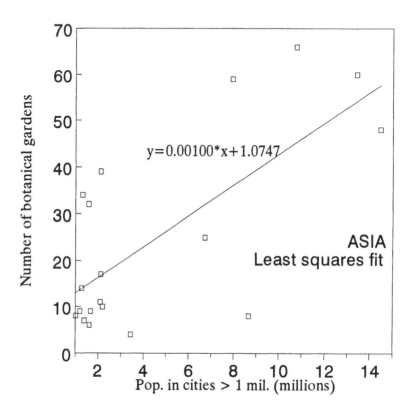

Figure 9.6. Asia, least squares fit. Botanical gardens and urban data.

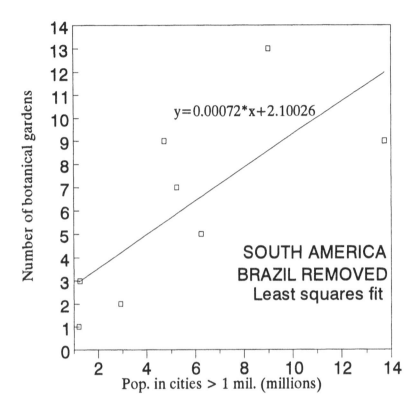

$$y=0.00072*x+2.10026$$

SOUTH AMERICA
BRAZIL REMOVED
Least squares fit

Figure 9.7. South America, least squares fit, Brazil data removed. Botanical gardens and urban data.

Residual plots: by continent

One might wonder further whether or not linear fits are the best fits for trying to capture any relation between people in cities of over 1 million and number of botanical gardens. Plotting the residuals (difference between the actual and estimated y-values: actual - regression estimate) can offer insight into this issue (Figures 9.8 and 9.9).

In any residual plot, the x-axis is the line of least squares for that scatter of values. Thus, if the residuals exhibit some pattern, it may be the case that a better fit is available; either a non-linear one for a clear non-linear pattern of residuals, or a linear one once outliers have been removed. If the linear plot shows no particular pattern, then the indication is that a linear fit was a reasonable choice. In the residual plot for Africa, there is a linear grouping of residuals in the center of the plot, but there are so many outliers to remove that it hardly seems useful to do so (Figure 9.8). The South America plot shows no particular pattern; a linear fit appears to have been a reasonable choice (Figure 9.8).

The residual plot for North and Central America does not have many values; what appears as a linear pattern with one outlier to one individual might not appear that way to another (Figure 9.9). Nonetheless, it will be investigated a bit further. The plot of residuals for Europe shows no strong pattern so that the regression line, the x-axis, does not describe any further association between the variables (Figure 9.9). The residual plot for Asia shows the most striking pattern of the set; there is a clear linear pattern except for a few outliers (Figure 9.9). Additional analysis of this situation , and of the North and Central America plot continues in the next section.

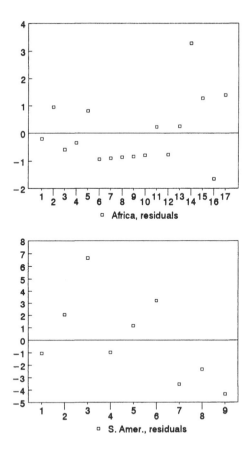

Figure 9.8. Africa and South America, residual plots.

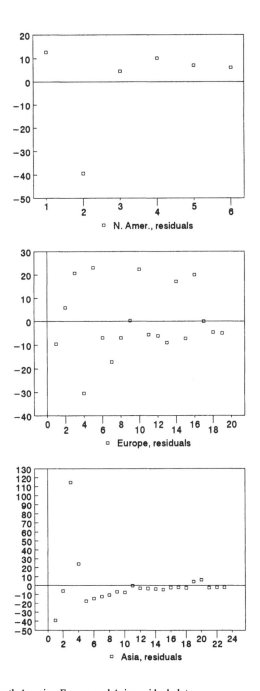

Figure 9.9. North America, Europe, and Asia, residual plots.

Additional analysis based on residual plots.

There is good reason that the Asia residual plot exhibits such a strong linear tendency. The four outliers to the left in Figure 9.9 represent the countries of China, India, the former U.S.S.R., and Japan. If these are discarded, all that remains is a large set of countries each of which has only a handful of botanical gardens. To remove the outliers in this case is therefore to remove any meaning from the association.

Thus, consider the remaining additional case: that of North and Central America. When the one outlier of Mexico is removed, and the regression performed on the remaining entries (albeit not many) the association improves, once again: the R-squared is nearly perfect (Table 9.4), the standard errors are negligible (Table 9.4), and the fit is visually nearly perfect (Figure 9.10).

TABLE 9.4

CENTRAL AND NORTH AMERICA, LESS MEXICO
Regression Output:

Constant	−0.94861969
Std Err of Y Est	2.462718697
R Squared	0.999605886
No. of Observations	5
Degrees of Freedom	3

X Coefficient(s)	0.002742596
Std Err of Coef.	0.000031441

Commentary

The notion that the number of botanical gardens might be a function of the number of people living in cities of 1 million or more was considered from a variety of perspectives. The results of the analyses are inconclusive; one might continue by investigating the relationship of other established variables to this new variable. There is a point to doing studies of this sort simply to build a volume of relationships that have been tried with a new variable. Even with established variables that have been tested in a number of ways, results of this sort are the rule. For every beautifully fit curve that gets published in scientific studies, there are usually reams of earlier inconclusive efforts, one of which opened the door to the eventual result!

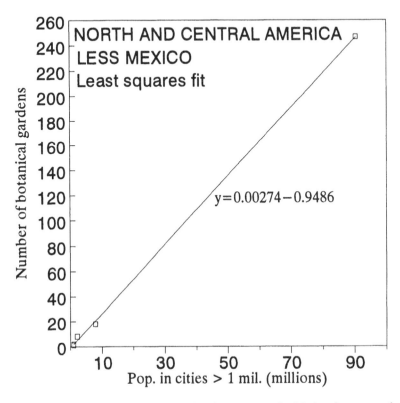

Figure 9.10. North and Central America, least squares fit, Mexico data removed. Botanical gardens and urban data.

CHAPTER 10

WORLD TRADE DATA ANALYSIS

ANALYTICAL TECHNIQUES/TOOLS USED:

Set-theoretic analysis of numerical structure of data
Partitioning of data
Organizational techniques
Geometric self-similarity

DATA TYPE: VARIABLE QUALITY OF DATA

cleaning, analysis, and graphing of data

Overview of Data

World trade is an unusually difficult concept to capture in data; on one level, the idea of an import/export is a simple matter. In a world in which economies of scale make it efficient to ship partially completed products from one country to be completed in another, it is often hard to know what is an export and what is an import. There are issues of how to count "ideas" as an item in the balance of trade and there is the problem of measuring trade in some consistent manner. In a world filled with myriad units of currency and variable rates of inflation this too causes considerable difficulty in assessing international trade.

With any data set (presented in electronic or paper format), it is important first to examine the set for interesting or unusual patterns in the display. These patterns often influence decisions in choosing subsets of data and tools to analyze subsets.

PATTERNS IN DATA--WHAT TO LOOK FOR

1. What is the general organizational scheme of the entire set? Is it arranged alphabetically, numerically, or in some other fashion?

2. Are the real-world entries in the Table (nations, states, counties) expressed as comparable units? For example, county data and national data are generally not comparable.

3.   Are the numerical entries in the Table expressed in comparable units? For example, data in one column might measure percentages while data in another column might measure thousands of dollars--these columns would not be comparable.

4. Are there gaps in the data? If so, what is their significance to the questions you wish to have the data answer?

Selection of related data sets: matching problems.

To illustrate some of the data problems, and associated difficulties with interpreting curves fit to world trade data, consider the data in Table 10.1, involving a mere single commodity (reptile skins) and all the variables in the World Resources Institute data base associated with reptile skins. One of the problems with looking at even just a single commodity, is that there are so many issues related to the trading of that item that are also useful to consider. Thus, Table 10.1 shows import and export data for reptile skins--one set for import, another for export. According to points 1 to 4 above, these two sets are ideal; they match perfectly, are arranged in the same order, and are measured in the same units over the same time interval. If all variables were compatible in this manner, the analysis of real-world data would be greatly simplified.

Perhaps this perfect meshing of data is too good to be true. Note that a country is either an importer or an exporter of reptile skins--not both. Practice tells us, however, that a reptile skin raised in Argentina might be shipped to Brazil where it is manufactured into a pair of shoes. The shoes might then be shipped back to Argentina where they are purchased. Yet Table 10.1 shows 0 imports for Argentina. Thus, one might suppose that once a skin is manufactured it is no longer counted as a skin; or one might suppose that the positive numbers in Table 10.1 represent the positive difference between import and export for a single country--that Argentina, for example, exports 1747153 more reptile skins than it imports. Prior to embarking on analyses that lead to policy recommendations it would be important to know why the columns in Table 10.1 are mutually exclusive.

Beyond the numerical structure of the entries in Table 10.1, it makes sense to ask questions related to the capability of nations to continue to meet the demand for reptile skins. Thus, it will be important to know that there is variety in the choice of reptiles and that their supply is adequate to sustain some sort of reasonable harvesting techniques. Thus, one might include data on number of known species of reptiles, numbers of reptile species threatened, and density of threatened species within a country (Table 10.2)

TABLE 10.1 (source: World Resources Institute)

Reptile skins

	Number Imported 1988	Number exported 1988
Argentina	0	1747153
Australia	691	0
Austria	96521	0
Belgium	71751	0
Bolivia	0	93708
Botswana	363	0
Brazil	8984	0
Cameroon	0	148510
Canada	119726	0
Chad	0	35710
Chile	3350	0
China	0	65665
Colombia	0	74173
Congo	0	1358
Zaire	0	8
Cyprus	0	1
Benin	0	3469
Denmark	546	0
Finland	2548	0
France	883971	0
Gabon	0	1
German Dem Re	14	0
Germany, Fed R	8381	0
Ghana	0	345
Greece	196	0
Guinea	0	18434
Guyana	0	72521
Honduras	0	15253
Hong Kong	267393	0
Hungary	25	0
India	0	3821
Indonesia	0	3032189
Israel	8064	0
Italy	426546	0
Cote d'Ivoire	0	46
Japan	950047	0
Kenya	0	1400
Korea, Rep	25199	0
Lebanon	346	0
Liberia	0	5
Madagascar	0	3177
Malawi	0	1830
Malaysia	0	238205
Mali	0	406312

Malta	1470	0
Mauritius	16	0
Mexico	106457	0
Morocco	342	0
Mozambique	0	795
Namibia	0	4
Netherlands	3444	0
New Zealand	3	0
Nigeria	0	2384
Pakistan	0	4
Panama	0	7470
Papua New Guin	0	23011
Paraguay	0	20
Peru	7	0
Philippines	0	35401
Portugal	1323	0
Guinea–Bissau	0	932
Saudi Arabia	8839	0
Senegal	0	9200
Sierra Leone	0	5
Singapore	445873	0
Viet Nam	0	1000
South Africa	14940	0
Zimbabwe	0	11607
Spain	625457	0
Sudan	0	106702
Sweden	1240	0
Switzerland	111570	0
Syrian Arab Rep	1	0
Thailand	0	260080
Togo	0	16415
United Arab Em	0	67750
Tunisia	0	399
Turkey	34566	0
Uganda	0	1
United Kingdom	583585	0
Tanzania	0	2318
United States	1641308	0
Uruguay	0	29838
Venezuela	0	92294
Zambia	0	3754
Taiwan	162888	0
WORLD	6634678	6634678
AFRICA	15661	775117
SOUTH AMER	12341	2378707
NORTH & CEN	1867491	22723
EUROPE	2818588	0
ASIA	1903216	3704116
OCEANIA	694	23011

Table 10.2 contains a great deal more data than does Table 10.1. Indeed, considerable merging of data will be required to make the two tables match. Within Table 10.2, there are also difficulties. Once again, it is the case that two of the columns match perfectly: the one showing the number of species threatened, and the number threatened per 10,000 kilometers squared (as one would hope). Surprisingly, perhaps, there is less data about the numbers of known species that there is about the number of threatened species. Evidently there is a serious gap in the reporting of information (at least). Thus, Angola has two threatened species, but does not list that it has any known species; it is difficult to imagine that one can know a species is threatened if one does not know it exists. Clearly there is a severe problem; Angola is not an isolated example; the data on species threatened is almost twice as long as the data on species known. Despite all these clear difficulties with this particular data set, it would no doubt still be prudent, for the sake of organization if nothing else, to systematically run through the check list of points 1 to 4 above for the data in Table 10.2.

1. What is the general organizational scheme of the entire set? Is it arranged alphabetically, numerically, or in some other fashion?

All the data is arranged alphabetically.

2. Are the real-world entries in the Table (nations, states, counties) expressed as comparable units? For example, county data and national data are generally not comparable.

Some data is expressed at a national level and some is at a continental level; in sorting sets, one needs to be careful not to accidentally mix the two levels.

3. Are the numerical entries in the Table expressed in comparable units? For example, data in one column might measure percentages while data in another column might measure thousands of dollars--these columns would not be comparable.

Some data is in absolute numbers; some is per 10,000 sq. km.

4. Are there gaps in the data? If so, what is their significance to the questions you wish to have the data answer?

There are many gaps, as noted above and in Table 10.2--note differences in column lengths in that table.

TABLE 10.2 (source World Resources Institute)

Reptile species

	Number known 1988		Number Threatened 1988	Number threatened Per 10,000 km sq 1988
Argentina	204	Afghanistan	1	0.251929202
Australia	550	Albania	1	0.705747762
Austria	13	Algeria	0	0
Bahamas	39	Angola	2	0.406844593
Bolivia	180	Argentina	4	0.625464502
Botswana	158	Australia	9	1.00448981
Brazil	467	Austria	0	0
Belize	107	Bahamas	3	2.69235753
Myanmar	360	Bahrain	0	0
Canada	42	Bangladesh	14	5.80588479
Chile	82	Barbados	0	0
Colombia	383	Belgium	0	0
Comoros	26	Bhutan	1	0.600078003
Costa Rica	218	Bolivia	4	0.848370452
Cuba	100	Botswana	1	0.261602517
Denmark	5	Brazil	11	1.18708757
Ecuador	345	Belize	3	2.28034487
El Salvador	92	Solomon Islands	3	2.11361054
Ethiopia	6	Brunei	3	3.59796954
Finland	5	Bulgaria	1	0.452022099
France	36	Myanmar	10	2.48886324
French Guiana	136	Burundi	1	0.713363102
German Dem Re	12	Cambodia	6	2.30720217
Germany, Fed R	12	Cameroon	2	0.559227357
Guatemala	204	Canada	0	0
Guyana	137	Cape Verde	1	1.34973756
Honduras	161	Cayman Islands	2	6.66943111
India	400	Central African l	2	0.511509308
Ireland	1	Sri Lanka	3	1.61257953
Italy	46	Chad	2	0.402905817
Jamaica	38	Chile	0	0
Japan	86	China	7	0.726095799
Kenya	191	Colombia	10	2.09584179
Luxembourg	8	Comoros	0	0
Madagascar	259	Congo	2	0.623449643
Malawi	124	Zaire	2	0.330261387
Mali	16	Costa Rica	2	1.16748443
Mauritius	19	Cuba	4	1.80835746
Mexico	717	Czechoslovakia	0	0
Mozambique	170	Benin	2	0.899491105
Netherlands	7	Denmark	0	0
New Zealand	39	Dominican Rep	4	2.37184976
Nicaragua	162	Ecuador	8	2.6528772
Nigeria	114	El Salvador	1	0.782338868

Norway	5	Equatorial Guine	2	1.42302376
Panama	212	Ethiopia	1	0.204775611
Paraguay	110	Fiji	4	3.27858366
Peru	297	Finland	0	0
Philippines	197	France	2	0.532500445
Portugal	35	French Guiana	2	0.968567668
Reunion	6	Djibouti	0	0
Viet Nam	180	Gabon	2	0.675961709
South Africa	301	Gambia, The	2	1.92094137
Zimbabwe	155	German Dem Re	0	0
Spain	64	Germany, Fed R	0	0
Suriname	131	Ghana	2	0.702157913
Sweden	6	Greece	3	1.2803913
Switzerland	15	Guatemala	4	1.81908902
Trinidad & Toba	76	Guinea	1	0.347594584
Turkey	93	Guyana	3	1.09002526
U.S.S.R.	144	Haiti	4	2.85616442
United Kingdom	11	Honduras	3	1.35133862
Tanzania	273	Hungary	0	0
United States	368	Iceland	0	0
Uruguay	66	India	17	2.51127159
Venezuela	246	Indonesia	13	2.29937795
Zambia	152	Iran, Islamic Rep	4	0.742102873
		Iraq	0	0
		Ireland	0	0
		Israel	1	0.785680467
		Italy	2	0.650091598
		Cote d'Ivoire	1	0.317835915
		Jamaica	3	2.90798402
		Japan	0	0
		Jordan	0	0
		Kenya	2	0.523609311
		Korea, DPR	0	0
		Korea, Rep	0	0
		Kuwait	0	0
		Lao PDR	5	1.75964088
		Lebanon	1	0.987140563
		Lesotho	0	0
		Liberia	2	0.902810254
		Libya	1	0.181559209
		Luxembourg	0	0
		Madagascar	10	2.60819261
		Malawi	1	0.442279831
		Malaysia	12	3.78599696
		Mali	2	0.407548108
		Mauritania	1	0.216964378
		Mauritius	6	10.4523027
		Mexico	16	2.8041882
		Mongolia	0	0
		Morocco	0	0
		Mozambique	1	0.235339997
		Oman	0	0
		Namibia	2	0.466362461

Nepal	9	3.76013669
Netherlands	0	0
Vanuatu	1	0.936739378
New Zealand	1	0.336608767
Nicaragua	2	0.85788424
Niger	1	0.202340919
Nigeria	2	0.449152606
Norway	0	0
Pakistan	6	1.41524599
Panama	2	1.01938733
Papua New Guin	1	0.282103067
Paraguay	4	1.17755688
Peru	6	1.2083387
Philippines	6	1.95299547
Poland	0	0
Portugal	0	0
Guinea−Bissau	2	1.30910126
Reunion	0	0
Romania	1	0.351585542
Rwanda	2	1.45287012
Sao Tome & Prir	0	0
Saudi Arabia	0	0
Senegal	2	0.748272846
Seychelles	2	6.50830422
Sierra Leone	2	1.04382752
Viet Nam	8	2.51911679
Somalia	1	0.253796467
South Africa	3	0.614469584
Zimbabwe	1	0.298356635
Spain	5	1.37071226
Sudan	1	0.161564937
Suriname	1	0.397869722
Swaziland	1	0.833582507
Sweden	0	0
Switzerland	0	0
Syrian Arab Rep	1	0.38167521
Thailand	9	2.4539757
Togo	2	1.12750968
Trinidad & Toba	0	0
United Arab Em	0	0
Tunisia	1	0.397596682
Turkey	5	1.1876265
Uganda	1	0.352380551
U.S.S.R.	3	0.235247672
Egypt	2	0.437343158
United Kingdom	0	0
Tanzania	3	0.668675035
United States	25	2.61352121
Burkina Faso	2	0.670606467
Uruguay	2	0.774225059
Venezuela	2	0.451049122
Western Samoa	0	0
Yugoslavia	1	0.343077962
Zambia	2	0.480575804

The matching of related data sets.

Tables 10.1 and 10.2 contain all the data within the World Resources Institute data base, for all available countries, concerning the trade of reptile skins. There is a real dilemma in further processing this data--in its present form, comparisons are meaningless because the sets are so different from each other. However, to reduce them to a set of common measures is to discard a great deal of information. To see exactly how much information gets lost, form the set-theoretic intersection of the three data sets (Table 10.3). What once took three pages to enumerate can now be enumerated on a single page; two-thirds of the entries of two of the variables have been removed. Table 10.3 is arranged alphabetically; that is useful for looking up values for particular countries, but the alternation of clumps of 0 entries obscures any other pattern in that table. In addition, the entries have been rank-ordered, for future graphing purposes, and a column calculating the percent of endangered species by country has been added.

When the data of Table 7.3 are sorted according to the column on percent of endangered reptile species, Table 7.4 emerges. From this table, one notes that countries with high levels of endangered reptile species may be either importers or exporters. Presumably the importers need to import skins to supplement their indigenous stock for local industry; on the other hand, export may be what threatens the reptile stock in the exporting countries. Note the set of countries at the bottom of the table that are all importers and have no threatened reptile species; they are all developed countries.

When the data are graphed on the percent threatened column, using the rank-ordering as the independent input, the sharply decreasing exponential-like graph of Figure 10.1 is produced. Graphs of this sort often emerge in situations of rank ordering.

Still, there are columns that are difficult to assimilate visually in Table 10.4. This difficulty can be removed, although the rank-ordering of the entire set will vanish, by separating the entries into two subsets-- one of exporters and one of importers. Table 10.5 shows the results of sorting the data in this manner. Each rank-ordered subset can also be graphed--there is no necessary reason that the graphs of the subsets should exhibit exponential decay--one might and the other might be relatively flat. However, when the percent threatened reptile species is graphed by subset--one for the exporters (Figure 10.2) and one for the importers (Figure 10.3), curves similar to the one in Figure 10.1, for the entire distribution emerge. The curves for the subsets are self-similar to the curve for the larger set. Thus, an exciting direction one might pursue in trying to establish order in this morass of missing data could involve concepts and tools from fractal geometry.

TABLE 10.3 (source: World Resources Institute)

Reptile skins			Reptile species			Percent
	Number Imported 1988	Number exported 1988	Number known 1988	Number Threatened 1988	# threatened Per 10,000 km sq 1988	threatened E5/D5*100
1 Argentina	0	1747153	204	4	0.625464502	1.960
2 Australia	691	0	550	9	1.00448981	1.636
3 Austria	96521	0	13	0	0	0
4 Bolivia	0	93708	180	4	0.848370452	2.222
5 Botswana	363	0	158	1	0.261602517	0.632
6 Brazil	8984	0	467	11	1.18708757	2.355
7 Canada	119726	0	42	0	0	0
8 Chile	3350	0	82	0	0	0
9 Colombia	0	74173	383	10	2.09584179	2.610
10 Denmark	546	0	5	0	0	0
11 Finland	2548	0	5	0	0	0
12 France	883971	0	36	2	0.532500445	5.555
13 German Dem l	14	0	12	0	0	0
14 Germany, Fed	8381	0	12	0	0	0
15 Guyana	0	72521	137	3	1.09002526	2.189
16 Honduras	0	15253	161	3	1.35133862	1.863
17 India	0	3821	400	17	2.51127159	4.25
18 Italy	426546	0	46	2	0.650091598	4.347
19 Japan	950047	0	86	0	0	0
20 Kenya	0	1400	191	2	0.523609311	1.047
21 Madagascar	0	3177	259	10	2.60819261	3.861
22 Malawi	0	1830	124	1	0.442279831	0.806
23 Mali	0	406312	16	2	0.407548108	12.5
24 Mauritius	16	0	19	6	10.4523027	31.57
25 Mexico	106457	0	717	16	2.8041882	2.231
26 Mozambique	0	795	170	1	0.235339997	0.588
27 Netherlands	3444	0	7	0	0	0
28 New Zealand	3	0	39	1	0.336608767	2.564
29 Nigeria	0	2384	114	2	0.449152606	1.754
30 Panama	0	7470	212	2	1.01938733	0.943
31 Paraguay	0	20	110	1	0.282103067	0.909
32 Peru	7	0	297	6	1.2083387	2.020
33 Philippines	0	35401	197	6	1.95299547	3.045
34 Portugal	1323	0	35	0	0	0
35 Viet Nam	0	1000	180	8	2.51911679	4.444
36 South Africa	14940	0	301	3	0.614469584	0.996
37 Zimbabwe	0	11607	155	1	0.298356635	0.645
38 Spain	625457	0	64	5	1.37071226	7.812
39 Sweden	1240	0	6	0	0	0
40 Switzerland	111570	0	15	0	0	0
41 Turkey	34566	0	93	5	1.1876265	5.376
42 United Kingdo	583585	0	11	0	0	0
43 Tanzania	0	2318	273	3	0.668675035	1.098
44 United States	1641308	0	368	25	2.61352121	6.793
45 Uruguay	0	29838	66	2	0.774225059	3.030
46 Venezuela	0	92294	246	2	0.451049122	0.813
47 Zambia	0	3754	152	2	0.480575804	1.315

TABLE 10.4 (source: World Resources Institute)

	Reptile skins		Reptile species			Percent
	Number Imported 1988	Number exported 1988	Number known 1988	Number Threatened 1988	# threatened Per 10,000 km sq 1988	threatened E5/D5*100
1 Mauritius	16	0	19	6	10.4523027	31.57
2 Mali	0	406312	16	2	0.407548108	12.5
3 Spain	625457	0	64	5	1.37071226	7.812
4 United States	1641308	0	368	25	2.61352121	6.793
5 France	883971	0	36	2	0.532500445	5.555
6 Turkey	34566	0	93	5	1.1876265	5.376
7 Viet Nam	0	1000	180	8	2.51911679	4.444
8 Italy	426546	0	46	2	0.650091598	4.347
9 India	0	3821	400	17	2.51127159	4.25
10 Madagascar	0	3177	259	10	2.60819261	3.861
11 Philippines	0	35401	197	6	1.95299547	3.045
12 Uruguay	0	29838	66	2	0.774225059	3.030
13 Colombia	0	74173	383	10	2.09584179	2.610
14 New Zealand	3	0	39	1	0.336608767	2.564
15 Brazil	8984	0	467	11	1.18708757	2.355
16 Mexico	106457	0	717	16	2.8041882	2.231
17 Bolivia	0	93708	180	4	0.848370452	2.222
18 Guyana	0	72521	137	3	1.09002526	2.189
19 Peru	7	0	297	6	1.2083387	2.020
20 Argentina	0	1747153	204	4	0.625464502	1.960
21 Honduras	0	15253	161	3	1.35133862	1.863
22 Nigeria	0	2384	114	2	0.449152606	1.754
23 Australia	691	0	550	9	1.00448981	1.636
24 Zambia	0	3754	152	2	0.480575804	1.315
25 Tanzania	0	2318	273	3	0.668675035	1.098
26 Kenya	0	1400	191	2	0.523609311	1.047
27 South Africa	14940	0	301	3	0.614469584	0.996
28 Panama	0	7470	212	2	1.01938733	0.943
29 Paraguay	0	20	110	1	0.282103067	0.909
30 Venezuela	0	92294	246	2	0.451049122	0.813
31 Malawi	0	1830	124	1	0.442279831	0.806
32 Zimbabwe	0	11607	155	1	0.298356635	0.645
33 Botswana	363	0	158	1	0.261602517	0.632
34 Mozambique	0	795	170	1	0.235339997	0.588
35 Austria	96521	0	13	0	0	0
36 Canada	119726	0	42	0	0	0
37 Chile	3350	0	82	0	0	0
38 Denmark	546	0	5	0	0	0
39 Finland	2548	0	5	0	0	0
40 German Dem l	14	0	12	0	0	0
41 Germany, Fed	8381	0	12	0	0	0
42 Japan	950047	0	86	0	0	0
43 Netherlands	3444	0	7	0	0	0
44 Portugal	1323	0	35	0	0	0
45 Sweden	1240	0	6	0	0	0
46 Switzerland	111570	0	15	0	0	0
47 United Kingdo	583585	0	11	0	0	0

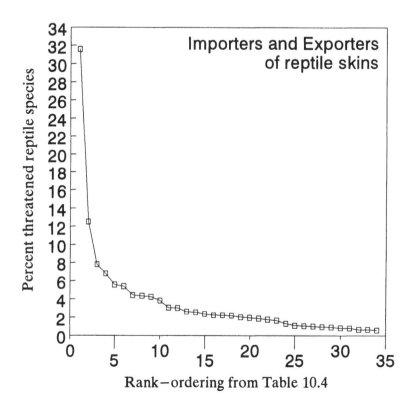

Figure 10.1. Importers and exporters of reptile skins.

TABLE 10.5 (source: World Resources Institute)

Reptile skins	Number Imported 1988	Number exported 1988	Reptile species Number known 1988	Number Threatened 1988	# threatened Per 10,000 km sq 1988	Percent threatened E5/D5*100
1 Mauritius	16	0	19	6	10.4523027	31.57
2 Spain	625457	0	64	5	1.37071226	7.812
3 United States	1641308	0	368	25	2.61352121	6.793
4 France	883971	0	36	2	0.532500445	5.555
5 Turkey	34566	0	93	5	1.1876265	5.376
6 Italy	426546	0	46	2	0.650091598	4.347
7 New Zealand	3	0	39	1	0.336608767	2.564
8 Brazil	8984	0	467	11	1.18708757	2.355
9 Mexico	106457	0	717	16	2.8041882	2.231
10 Peru	7	0	297	6	1.2083387	2.020
11 Australia	691	0	550	9	1.00448981	1.636
12 South Africa	14940	0	301	3	0.614469584	0.996
13 Botswana	363	0	158	1	0.261602517	0.632
14 Austria	96521	0	13	0	0	0
15 Canada	119726	0	42	0	0	0
16 Chile	3350	0	82	0	0	0
17 Denmark	546	0	5	0	0	0
18 Finland	2548	0	5	0	0	0
19 German Dem l	14	0	12	0	0	0
20 Germany, Fed	8381	0	12	0	0	0
21 Japan	950047	0	86	0	0	0
22 Netherlands	3444	0	7	0	0	0
23 Portugal	1323	0	35	0	0	0
24 Sweden	1240	0	6	0	0	0
25 Switzerland	111570	0	15	0	0	0
26 United Kingdo	583585	0	11	0	0	0
27 Mali	0	406312	16	2	0.407548108	12.5
28 Viet Nam	0	1000	180	8	2.51911679	4.444
29 India	0	3821	400	17	2.51127159	4.25
30 Madagascar	0	3177	259	10	2.60819261	3.861
31 Philippines	0	35401	197	6	1.95299547	3.045
32 Uruguay	0	29838	66	2	0.774225059	3.030
33 Colombia	0	74173	383	10	2.09584179	2.610
34 Bolivia	0	93708	180	4	0.848370452	2.222
35 Guyana	0	72521	137	3	1.09002526	2.189
36 Argentina	0	1747153	204	4	0.625464502	1.960
37 Honduras	0	15253	161	3	1.35133862	1.863
38 Nigeria	0	2384	114	2	0.449152606	1.754
39 Zambia	0	3754	152	2	0.480575804	1.315
40 Tanzania	0	2318	273	3	0.668675035	1.098
41 Kenya	0	1400	191	2	0.523609311	1.047
42 Panama	0	7470	212	2	1.01938733	0.943
43 Paraguay	0	20	110	1	0.282103067	0.909
44 Venezuela	0	92294	246	2	0.451049122	0.813
45 Malawi	0	1830	124	1	0.442279831	0.806
46 Zimbabwe	0	11607	155	1	0.298356635	0.645
47 Mozambique	0	795	170	1	0.235339997	0.588

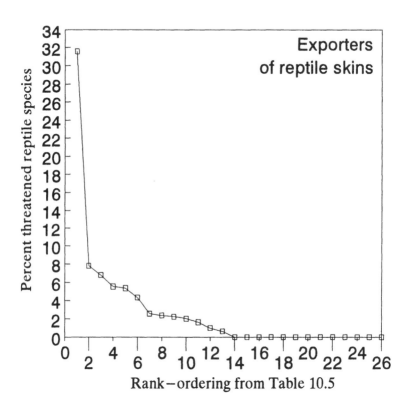

Figure 10.2. Exporters of reptile skins.

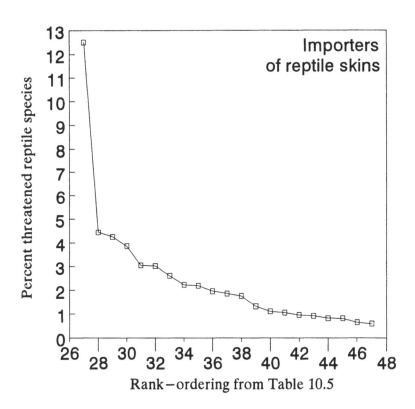

Figure 10.3. Importers of reptile skins.

INDEX OF FIGURE CAPTIONS
AND TABLE TITLES

FIGURE CAPTIONS

TABLE TITLES